Principles of
Construction

Second Edition

ROGER GREENO

Longman

An imprint of **Pearson Education**

Harlow, England · London · New York · Reading, Massachusetts · San Francisco
Toronto · Don Mills, Ontario · Sydney · Tokyo · Singapore · Hong Kong · Seoul
Taipei · Cape Town · Madrid · Mexico City · Amsterdam · Munich · Paris · Milan

Pearson Education Limited
Edinburgh Gate
Harlow
Essex CM20 2JE
England

and Associated Companies throughout the world

Visit us on the World Wide Web at:
http://www.pearsoneduc.com

First published 1986
British Library Cataloguing in Publication Data
A catalogue entry for this title is available from the British Library.

ISBN 0-582-23086-1

Library of Congress Cataloging-in-Publication data
A catalog entry for this title is available from the Library of Congress.

10 9 8 7 6 5
06 05 04 03 02

Set by 4 in $10^1/_2/12$pt Times
Printed in Malaysia, PP

CONTENTS

ACKNOWLEDGEMENTS

We are indebted to the following for permission to reproduce copyright material:

British Standards Institution for extracts from BS3921: 1974. Complete copies of the document can be obtained from BSI at Linford Wood, Milton Keynes, MK14 6LE; HMSO for extracts from the Building Regulations by permission of the controller of HMSO Crown Copyright.

Authors acknowledgements are extended to the series editor Mr. C.R. Bassett for his help and encouragement during the preparation of this book.

PREFACE

This book is intended as guidance for students pursuing any elementary building course. Its illustrative text commends its application as a useful preliminary reader for degree, diploma and professional studies in building and a complementary reference to all first and second-year BTEC construction courses.

There are ideas and suggestions which will appeal to practical householders with intentions to maintain or improve their property by enlargement or general refurbishment.

The content acknowledges the processes involved at preconstruction stage in addition to the principles and controls applied throughout the construction programme. These include most aspects of elementary construction with an insight into techniques applied in larger-scale development using standard steel sections and reinforced concrete.

Superficial and structural defects are revealed, with details of remedial treatment plus guidance for the procedure undertaken in a structural survey. Current legislation and standards receive appropriate reference, with particular attention to safety and correct use of materials.

ROGER GREENO

1

The building cycle

Duties and relationship between the various parties to a building contract

Most construction work, regardless of scale, is obtained by competition between contractors for submission of the best price. A client provides the financial resources to fulfil his objectives and he approaches several builders, either directly or through the professional capacity of an architect. Small works, i.e. maintenance, repairs, alterations and extensions on a domestic housing level, are normally undertaken without the need for an architect. Extensions to houses are unlikely to justify an architect; most building contractors are capable of preparing and submitting their own plans.

The usual procedure is for the client to present a brief to an architect, containing in comprehensive terms the size, capacity, style of construction and building function that is required. The architect will consider the feasibility and prepare an initial design for client approval. When the design is complete, professional consultants are engaged for specialist installations such as structural steelwork, services, lifts, etc. They are involved in the detailed design which is finally sent to the local planning and building control authorities for consideration.

A quantity surveyor (often working in partnership with an architect) uses the drawings to compile a bill of quantities. This is a complete list of all materials measured net, and accompanies the design drawings when sent to several building contractors to price. Competitive estimating between builders is known as tendering, and usually the lowest estimate secures the contract. Lack of confidence in the builder's work may deter an architect from accepting the lowest price. An outline of the client/architect/builder relationship is shown in Fig. 1.1, including other interested parties at design and construction stages.

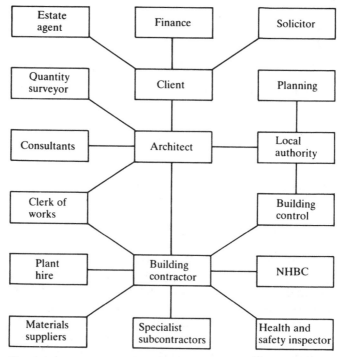

Fig. 1.1 Associates involved in a building contract

Occasionally the architect and builder modules combine, to form a design and construct partnership to reduce communication difficulties and offer a more efficient and economic service to the client.

Speculative building contractors have no client to finance the work and build specifically with the objective of selling to an unknown market. This can be lucrative when housing funds are readily available and properties scarce, but otherwise it has proved the downfall of many good builders. A design consultancy or architect are usually involved initially and estate agents are used to market the buildings on completion.

Roles and responsibilities summarised

Client

1   Present the architect with a comprehensive account of his requirements with particular regard to use, space (floor area and volume), cost limit and completion date.

2   Arrange finance.
3   Appoint estate agent (if speculative venture).
4   Employ solicitor to prepare contracts of land purchase and to review legal commitments with architect and builder.
5   Liaise with architect during preparation of preliminary designs and feasibility study.
6   During contract, honour completion of interim valuations by stage payments to the contractor.
7   On completion, settle all agreed financial claims and professional fees.
8   Appoint a "Planning Supervisor" and nominate the "Principal Contractor" – CDM Regs.

Architect

1   Initially advise the client on the feasibility of proposals and approximate costs.
2   Investigate the site if it exists at this stage, or advise on the location of possible sites.
3   Prepare sketches and models to illustrate his ideas.
4   Provide a team of specialists to advise on specialist services such as heating and air-conditioning, electrical installations, structural requirements, etc.
5   Obtain agreement between client, consultants and himself on size, shape, layout and content.
6   Produce detailed drawings and materials specifications for local authority and quantity surveyor.
7   Select suitable building contractors, provide them with details and bill of quantities to price (work put out to tender).
8,   Check tenders and select the most suitable.
9   During construction, issue necessary instructions and variation orders to the builder and inform client of progress.
10   Provide interim certificates of completion for the builder to receive payment as the contract proceeds.
11   Certify completion for full payment from the client to the builder, less a retention which is held to ensure defects are made good within an agreed period.
12   Certify release of retention, (a small percentage of the contract price to cover defects and repairs occurring within a short period after completion).
13   If required, function as "Planning Supervisor" – CDM Regs.

4 *Principles of construction*

Consultants (generally structural and services)

1. Advise the designer of the most efficient, economical and practical method of providing their specialist service.
2. Prepare detailed drawings within the architect's design.
3. Prepare design calculations and material specifications.
4. During construction, check quality of workmanship and correct use of materials.

Quantity surveyor

1. At pre-contract stage, advise architect on costs and financial viability of certain construction techniques and materials.
2. Prepare approximate costs from architect's initial design brief and sketches.
3. Prepare bill of quantities by measuring and listing the quantities of all materials used in the building in accordance with the current Standard Method of Measurement.
4. Check builder's priced bill of quantities and advise architect of errors and accuracy of estimates.
5. During construction, prepare stage or period valuations.
6. Agree the cost of amendments and variation orders with the builder and architect.
7. Prepare the final account.
8. Advise architect of additional costs.

Clerk of works

Generally defined as the architect's representative on the site. His function is to ensure compliance with the architect's drawings, specifications and instructions by checking and inspecting all aspects of construction. The clerk of works reports direct to the architect, as he has no authority over the contractor. He can of course advise and comment on the work to the contractor and assist with difficulties in interpreting the architect's design.

Builder/contractor

1. Consider whether he has sufficient resources, (labour, plant and equipment) and time with regard to other commitments to fulfil the contractual obligations.
2. Conduct a site visit to anticipate difficulties not revealed on the drawings or in the bill of quantities.
3. Obtain and co-ordinate subcontractors and suppliers (plant and materials) prices and produce an estimate based on these, the drawings, the bill of quantities and the outcome of site investigations.

4. Agree the estimate with the board of directors or their appointee and submit it to the architect.
5. If the estimate is accepted, the builder will be given the opportunity to meet with the client, architect and quantity surveyor to agree alterations and amendments.
6. Hold preconstruction meetings to consider allocation of staff and develop contract programme.
7. Co-ordinate programme with architect, subcontractors and suppliers.
8. During construction, the contractor's roles and responsibilities are considerable. Briefly, they include management of direct staff, subcontractors, plant and materials during the contracted period of construction with due regard to the relevant codes of quality, safety and legislation. To fulfil these obligations the builder will appoint a manager, generally known as the agent or 'general' foreman, who will also possess the qualities required to harmonise co-ordination between the numerous different agencies on site.
9. At interim stages, and at contract completion, provide the quantity surveyor with sufficient data to complete his valuations and accounts.
10. After completion, undertake remedial work within the contracted period.
11. If nominated, function as "Principal Contractor" – CDM Regs.

Local authority

Applications for constructional proposals are received by the appropriate local authority. They are considered by the planning, building control and highways departments. The planning department's prime function is to ensure that the proposal is suited to the site and does not breach government development restrictions such as 'Green Belt' policy. They must also ensure that the design and materials are aesthetically acceptable and that the proposal will not impose excessively on local facilities. These include roads, drains, shops, schools, etc. Most planning legislation is contained in the Town and Country Planning Acts.

The building control section appoint building control officers to particular areas of jurisdiction. Their function is to ensure that the proposal complies with the Building Regulations, with particular regard to suitability and strength of materials. There are specific states of construction when the official must be given notice to visit the site, although he may call at any time.

Subcontractors

These are usually employed to provide a specialised or supplementary service. Where the construction requires specialised installations and techniques not undertaken by the main contractor such as lifts, escalators, air-conditioning, fire prevention, etc., the architect prepares a list of suitable contractors agreeable to the main contractor and asks them to submit tenders for the work. Alternatively the architect may nominate particular subcontractors known to have the sufficient resources to undertake the work efficiently. Labour-only sub-contractors are another possibility, generally employed by the main contractor to supplement his own workforce, possibly for the whole contract or at significant stages.

Suppliers – Builders' merchants, timber merchants, manufacturers and plant hire

These are suppliers of general building materials, equipment or specialised materials. If the last they may be nominated by the architect because of their particular style or quality or product. Suppliers responsibilities include:

1 Preparation of quotations.
2 Ensuring material quality meets architects specifications.
3 Delivery of materials in accordance with the contract programme.
4 Submission of accounts at regular intervals.
5 Provision of agreeable trade and bulk discounts, plus additional 5% if payment is received within 30 days of invoicing.

National House Building Council

This is a non-profit making private organisation created to form a register of house builders capable of producing high standards of construction. The NHBC have their own rules and regulations which supplement the Building Regulations and these are enforced by a team of inspectors distributed throughout the country. Inspectors may examine members' housing projects at any constructional stage, and will issue a ten-year structural warranty to the purchaser on satisfactory completion of the dwelling. The Council's prime objective is to protect the purchaser from disreputable builders and to provide improved standards of accommodation in modern housing. The builder is also protected from undue harassment and unreasonable demands from dissatisfied clients. This organisation has been particularly useful in filling the gap which would normally be occupied by the clerk of works in larger-scale architect-controlled construction. Its influence can be appreciated by the fact that very few mortgagers will finance the purchase of non-NHBC warranted new housing.

Estate agents

Estate agents will market properties for a percentage free charged to the seller (vendor). This is usually about 2% but may be negotiated, particularly for speculative estate developments. If more than one agency is employed by the builder, the selling fee is likely to be 0.5% more.

Standard form of contract

The business relationship between client and builder can be entirely verbal, and often is for small alterations, repairs and maintenance. Settlement on satisfactory completion of the work is by payment by cheque or cash from client to builder. This process is simple and effective for small-scale contracts but complicates as the construction process enlarges. For example it would be entirely impractical for a client to hand over the total payment for a large development upon completion; the builder needs interim or stage payments to pay operatives and buy materials, and the client needs to retain a residual after completion to cover payment for immediate defects. There are also other financial complications to be agreed between the two parties, such as payment for uncontracted alterations (client variations in the specification) and unforeseen delays, e.g. non-billed excavation due to subsoil complications.

To simplify the contractual relationship between client and builder, a copy of the JCT Standard Form of Building Contract is issued with the drawings, bill of quantities, material specifications and tender form (shows builder's estimated price for the work), when the builder, is invited to tender. The Joint Contracts Tribunal is composed of many professional and trade organisations. They combined to form a standard form of building contract during the 1960s in order to rationalise the various other standard forms which had existed since the turn of the century. The new document endeavours to clarify terms of contract more precisely, leading to a reduction in the number of claims for

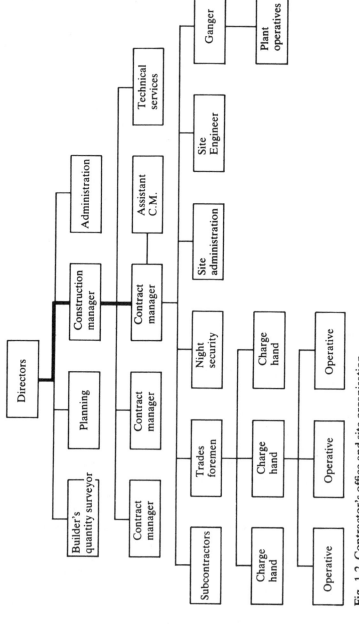

Fig. 1.2 Contractor's office and site organisation

compensation from both dissatisfied customers and builders. Variations exist to cover the diverse nature of the construction industry, and most contracts are divided into three distinct sections:

1. Articles of agreement. This section contains brief details of the contract:

(a)   date and duration;
(b)   names and addresses of all parties involved;
(c)   address and location of site;
(d)   contract sum;
(e)   quantity of drawings, their reference numbers and titles;
(f)   witnesses to the contract.

2. Contract conditions. This section is very involved and could contain up to 35 clauses. It details unambiguously the mutual liabilities, responsibility and obligations of client, builder and other contract-related parties.

3. Appendix. Any additional information not contained within the contract clauses may be defined in the appendix, eg. penalty clause details for failure to complete on time or amount of retention at stage payments – usually 5%.

Site personnel organisation

The administration and organisation of site staff will depend on the nature of work and scale of construction. Figure 1.2 shows the possible hierarchy of responsibility in a medium-sized firm, undertaking modest housing estate and light industrial types of development. The contract or site manager has overall responsibility and control on site, usually with the help of an assistant. They direct operations through their trades foremen and gangers, who in turn interpret these requirements by forming small groups of operatives and apprentices to fulfill the construction objectives. The contract manager's principle role is overall coordination of the various trades, with particular attention to programming material and plant deliveries to arrive at the appropriate stage of construction.

Functions and responsibilities

The following is a summary of the function and responsibilities of some of the more prominent personnel on site.

Contract manager

Otherwise known as the site agent because he is the builder's official or contact on site. Some organisations also refer to this person as the site supervisor, controller or general foreman, terms sometimes dictated by the span of responsibility and type and size of contract. His duties are too diverse to define, but they could include the following functions:

1. Controls the contract programme.
2. Organises site administration.
3. Engages supplementary labour.
4. Chases and checks material deliveries.
5. Co-ordinates subcontractors' work with his own direct staff.
6. Ensures health and safety provisions are adequate.
7. Considers clerk of works', safety officials' and inspectors' requests.
8. Organises site meetings with client and architect or their representatives.

Engineer

1. Sets out the initial excavation for foundations.
2. Locates and checks horizontal levels and plumbness of the structure.
3. Checks quality and suitability of materials, strength of concrete etc.

Builder's quantity surveyor

1. Measures work completed at specific stages.
2. Values architect's variations.
3. Values and pays for subcontractors' work.
4. Assesses bonus incentive values.
5. Schedules and monitors materials and plant.
6. Processes site cost control information for head office.

Site administration

This can be quite extensive, involving employment of numerous personnel. The timekeeper, storeman and canteen supervisor occur most frequently and their functions are briefly considered.

Timekeeper

1. Records attendance of site personnel.

2. Collects cash from bank or co-ordinates collection from a security firm.
3. Prepares wage slips and pay packets.
4. Distributes pay.

Storeman

1. Controls quantities of common materials, e.g. cement, bricks, sand, etc.
2. Distributes and registers allocation of materials and equipment.
3. Records delivery and return of hired equipment.

Canteen supervisor

1. Organises weekly menu.
2. Arranges delivery of food and preparation.
3. Prepares financial balance.

Canteen facilities are frequently undertaken by catering contractors to simplify the builder's administration.

Foreman and ganger

Responsible for organising skilled and semi-skilled operatives, respectively. These are rarely entirely administrative and work with their trades operatives. Charge hands are sometimes required to supervise small groups of operatives when the labour content is considerable or widely dispersed.

Operatives

These exist to fulfil the numerous trades in the building industry. Some are better qualified and experienced than others and, to acknowledge this differential, slight variations in pay are effected in some of the trades. Assistants, e.g. plumber's mate, exist where craftsmen require semiskilled help and apprentices are employed from school-leaving age for a period of three to four years training.

Control of resources

The major resources on site are materials, personnel and plant. Ultimate responsibility for their control lies with the site manager but he will delegate the responsibility to resource managers to cover each section. Materials are likely to be supervised by the storeman; personnel, the assistant site manager; and plant, possibly the storeman or a ganger on small developments, and a plant manager on larger sites.

Materials control

Material waste is a topical issue, as there have been many recent publications drawing attention to excessive waste and loss of materials on site. In the past, estimators have been content to add a small percentage (about 5%) to the contract price to cover loss and damage, but now the high cost of materials and competition for work justifies stricter site control.

Waste can be identified in several areas, conveniently dividing in two categories:

1. Direct waste. This is attributed mainly to loss due to irrepairable damage or theft from the site. Examples include:

(a) *Double handlings*; basically bad organisation as a result of management failing to design an effective site layout.

(b) *Transportation waste*; damage during loading, unloading and stacking on site. Generally bad handling.

(c) *Theft and vandalism*; theft is either directly from the site by site employees or trespassers, or by delivery drivers failing to unload the full amount – bad checking procedure.

(d) *Misuse*; using materials of a standard in excess of the specification, or below the specification and having to remove them and rebuild at a later date.

(e) *Specification errors*; the builder's costs incurred by this can be claimed against the client.

(f) *Poor management of plant*; plant lying idle or left running when not in use. Using equipment with a capacity in excess of that required.

(g) *Trainees' waste*; attributed to apprentices learning a trade, should be minimal if the trainee is well supervised.

(h) *Residual waste*; surplus mortar or concrete are usual examples, also paint left in open tins.

(i) *Dimensionally uncoordinated units*; waste attributed to time delays and material waste from cutting to fit awkward dimensions. Many building elements are now dimensionally co-ordinated, particularly factory prefabricated units.

(j) *Site storage*; damage by bad stacking, particularly bricks or blocks stacked too high and cement bags placed on damp ground or left uncovered. Careless site handling and transport will add to this.

(k) *Conversion waste*; waste left after cutting materials (particularly timber, sometimes bricks and blocks) to uneconomical shapes. The regularity of contemporary construction has reduced this.

2. *Indirect waste*. This is where materials are used for purposes other than that specified. Some possibilities include:

(a) *Substitution*; use of facing bricks below ground level or for partitions that will be plastered. Also employing bricks in block, walls where the bricklayer cannot be bothered to cut blocks, and using planed timber where sawn is adequate.

(b) *Production waste*; bad supervision of excavation work can result in over-digging, wasting time and consuming too much material. Also failure to account for intangibles at estimating stage, e.g. extra plant costs, delays due to repairs, insufficient trench timbering, etc.

The site manager will have enough problems chasing suppliers for deliveries and co-ordinating invoices and submitting returns to head office. Therefore he should appoint a responsible person to record and check all deliveries and to ensure that loads are stacked or stored in the correct position.

Site security is essential and includes fencing or hoarding with lockable gates, compounds and lockable sheds for extremely valuable and easily-marketed goods (e.g. shower mixers, basins, etc.) and illumination at night to deter thieves. The site manager's hut should be close to the site access with a good view of the site, and into vehicles as they arrive and leave. Spot checks may be implemented to deter theft by site operatives.

Personnel control

Control of site personnel is simplified by compartmentation of trades. Each trade is supervised by a foreman, who will subdivide his operatives into groups of about four under the leadership of a charge hand. Skilled craftsmen are responsible for an assistant or an apprentice.

The contract manager will keep a record of all operatives' details; pay number, insurance number, date employed, etc., which is supplemented with information relating to adverse timekeeping, misconduct, absence and other related information. Timekeeping techniques vary depending on the nature and scale of work. Construction over several months or more will justify a time clock registering arrival time in the morning, period taken for lunch and time of leaving in the afternoon/evening. Smaller sites may operate a signing on and off routine in an attendance book, or possibly just an acknowledgement to the foreman to signify arrival and departure. Non-productive time due to adverse weather is recorded by the site manager and operatives receive a nominal payment to cover this.

Productivity bonuses are difficult to establish in the building industry as the work does not relate to factory routine production and piece-rates. Also, inclement, weather is unaccountable and will affect productivity considerably. Output targets should be established in realistic terms otherwise disputes between management and operatives cause unnecessary delays, cancelling the efficiency which they are designed to encourage. Trades foremen must be aware of operatives' needs, ensuring adequate supply of materials and supplementary labour as required.

Subcontractors are largely responsible for their own organisation. If they are to fulfil their commitments efficiently the site manager must ensure co-ordination of the work programme (including plant and materials) with other subcontractors and the directly employed labour.

Plant control

Building plant and equipment is expensive to buy and hire, therefore careful planning is essential to ensure minimum idle time by deployment or transfer to another site. Plant utilisation sheets are often required by the contractor's head office to apply close control on use and maintenance. This ensures correct application, regular servicing and assessment of effectiveness.

2

Site preparation

Selection of a suitable site for development will be undertaken by the client with the help of his architect. Their search for a site of suitable potential is likely to be extensive, following up land advertisements in the local and national Press, estate agents' particulars and visits to auctions. Preliminary site investigations will be very detailed to ascertain the many physical aspects such as subsoil composition, demolition, etc., and the legal aspects which include planning permission, rights of access and preservation orders. Both physical and legal considerations are summarised in the following site investigation and information report and detailed in separate sections later in this chapter. The client's report could be available for builders' reference when invited to tender, although each builder will conduct his own site investigation to reinforce the client's information.

Site investigation

A site-visit report is divided into several sections and where necessary tabulated for easy reference. It should contain the following data:

Site investigation

(a) Location
- Nearest town or city
- Proximity of schools, emergency services, entertainment, recreation, shops, transport and employment, (all good selling points for speculative developments)
- Distance from head office (client and builder) and travelling time
- Distance from nearest railway station to the site

(b) *Accessibility*

- Approach and site access roads: width, gradient, bends, sharp corners, condition and construction relative to transport of heavy plant and equipment
- Bridges: strength, width and clearance height
- Temporary roads: rolled metal tracks or consider preparing the sub-base for new roads as a temporary access

(c) *Availability of space*

- Site offices, canteen, stores and compound
- Material storage areas and handling
- Construction area and assembly areas
- Plant location

(d) *Services*

- Water, drainage, electricity, gas and telephone
- Location will be determined from maps by consultation with the appropriate local authorities. Distance from site will affect cost of installation as will diversion of services already on site. Electricity potential (230V or 400V) should be established for operation of plant and equipment
- An estimate of building usage is also helpful for ascertaining the demand on sewers and drains

(e) *Ground composition*

Boreholes are required to determine:
- Changes in strata
- Strength of subsoils
- Toxicity of subsoils
- Stability of excavation
- Water table (depth below surface)

(f) *Site clearance and demolition*

A plan of the site should indicate trees, shrubs and existing buildings, and a site survey will reveal the extent of levelling necessary.

Demolition and excavation:

- Methods and costs
- Effect of tree and structural preservation orders
- Reuse of materials

- Protection of adjacent buildings
- Protection of other property and people
- Special insurance requirements
- Compensation payments and liability for damage
- Distance to spoil tips and charges

Local conditions

Local authority (planning, highways and building control departments)

Permission to develop land will be sought at an early stage by the client's architect. To ascertain the viability of a proposal, outline approval is sought to establish whether the nature of the development is acceptable. The highways department must also be consulted to ensure existing access roads and facilities are adequate and proposed alterations and improvements are acceptable. The builder must also satisfy this department when providing perimeter fences or hoardings to safeguard users of adjacent roads and footpaths.

Following outline approval, a detailed planning application is necessary to show proposed elevations and materials specifications. This is also referred to the building control department, for approval of structural and environmental factors.

Applications to the local authority attract a fee for both outline and detailed plans, plus an additional fee for building control. The amount varies from about £50 upwards, depending on the proposed scale and volume of development.

Labour availability

Enquiries at the local employment centre will reveal the availability of local skilled and unskilled labour. This is an important consideration for the builder, as most development work involves the temporary employment of labour for the contract duration only. An absence of labour due to several other large contracts in the area sets a high premium for craftsmen, or the prospect of high transportation costs. The quantity employed considerably affects the contractor's obligations for health and welfare facilities with regard to provision of lavatories, wash rooms, canteens and first-aid treatment.

Local resources

Builders' merchants. The builder will need to establish a list of local suppliers to ensure availability of materials. Negotiation will determine terms of business with particular regard to trade discounts, extent of credit, delivery charges and additional discount for prompt settlement of accounts.

Subcontractor (see also Ch. 1). Specialist subcontractors will be requested to tender for installation work not normally undertaken by the main contractors. The arrangement may be on a 'supply and fit' basis where the subcontractor provides all materials and fittings, plus the associated labour. An alternative is 'labour only', where the builder or client prefers to retain complete control over supply of equipment. Labour-only subcontractors are often employed where it is more convenient to hire and control bulk supplementary labour instead of a collection of individuals.

Plant hire. Very few builders can afford to purchase and maintain construction plant and equipment. Most large towns contain several plant hire businesses, which should enable the builder to obtain his requirements by competitive negotiation.

Climatic and weather conditions

Local weather conditions could affect the contract duration and advance preparations are better than confusion and costly remedial measures later. Structural design, style and materials should be compatible with local conditions, i.e. flat roofs should be avoided in areas of persistent rainfall. Snow and wind are variable imposed loads incorporated into the design, although the wind effect around a series of buildings (particularly of variable height) is often overlooked and human comfort seriously affected. Tall buildings should be discouraged where the possibility of wind damage, humidity or fog are likely. Solar radiation can be unpleasant, but if used sensibly and controlled can have many advantages including water heating through solar panels. Local industries must be treated with suspicion as atmospheric and noise pollution may detract from the value of any proposed nearby development. Orientation and construction of buildings should receive special design consideration where environmental and climatic factors have a significant effect.

Security

Security requirements vary depending upon the nature and size of the work, and the locality. The builder should negotiate with the local authority to ascertain minimum fencing standards, and the police for guidance on security measures in an unfamiliar area. The three areas of importance are theft of materials and equipment, vandalism and trespass particularly by children using the site as an adventure playground. Consultation with the authorities will determine whether fences, hoardings or compounds are used, and whether security patrols visit the site, or are permanently on watch during the non-productive hours. Floodlighting is often recommended by the police, and may be considered a useful night-time deterrent for thieves.

Site organisation

Temporary services

Connection of services to a building site may be temporary where the work is transient, e.g. highways. Elsewhere the services will be a permanent necessity and should be installed accordingly to avoid repeating the work. The relevant authorities operate on a national, regional and area basis for dispersion of responsibilities.

Water and sewage

There are numerous regional authorities, each independent of the other, generally located to administer common water sources. Their influence extends to all water resources, including lakes, ponds, rivers and some coastal areas. Each region has several subdivisions for convenience of local administration and these will process applications for connections to existing water mains and sewers.

Gas

The British Gas Corporation direct and maintain supplies nationally through regional boards, each subdivided into local area offices and showrooms. Applications for gas installations should be made to the appropriate area office.

Electricity

National Power, PowerGen and Nuclear Electric are the UK's national authorities for generation and primary distribution. Secondary distribution is controlled by regional boards subdivided into smaller administration areas to implement local service requirements.

Telephone

British Telecom is the national authority, controlling eleven regions each subdivided into sixty two areas for local installations. Temporary pay phones are often used in the site manager's office to control use.

Location and co-ordination

All services must be preplanned and located on a site plan in a representative colour scheme. To avoid repeated excavation of service trenches and possible damage to services already located, the combined service trench shown in Fig. 2.1 rationalises installation, provided co-ordination between the various authorities can be obtained. Foul water and surface water sewers are omitted as these are laid to a gradient which is incompatible with other services.

Construction (Design & Management) Regulations – 1994

These are widely applied to all construction work, with the exception of very small works. The client, architect and main contractor are obliged to provide a coordinated and integrated approach to health and safety issues throughout the duration of a building contract.

Client

The client's prime obligations are to appoint the "Planning Super-

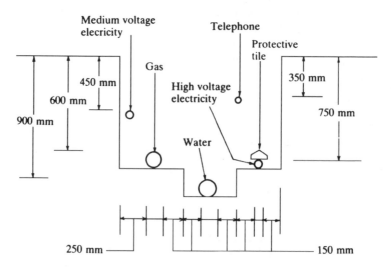

Fig. 2.1 Common services trench

visor" and "Principle Contractor". Other functions include ensuring that the workforce are competent and adequately resourced; providing the planning supervisor with relevant information about the land and/or premises; ensuring the principle contractor has prepared a health and safety plan to indicate perceived risks; and retention of a health and safety file for reference and inspection.

Planning supervisor

The planning supervisor must notify the HSE of the project; prepare a health and safety file; ensure designers comply with their duties, cooperating for purposes of health and safety; prepare a pre-tender health and safety plan; advise client on competence and resource allocation of all parties to the contract and implications of the health and safety plan before construction work begins.

Principal contractor

This is usually the main contractor, who coordinates and manages health and safety during construction work; effects the health and safety plan; ensures only competent and adequately resourced sub-contractors are engaged; coordinates and ensures cooperation of sub-contractors; assesses risks and determines policy; ensures sub-contractors are consulted and informed of identified risks; admits only authorized people to the site; monitors health and safety issues and conveys health and safety information to the planning supervisor for registration in the file.

Construction (Health, Safety & Welfare) Regulations – 1996

These set objectives, without dwelling on many specific or prescriptive requirements, effectively implementing assessment and management of risk. Areas of work include: potential for falls and falling objects; site transport, roads and traffic routes; segregation of people and traffic; ventilation, light and general safety in confined spaces; use of mechanical plant; emergency and fire procedures; excavations; electricity on site; assessment of welfare of personnel – personal protective equipment, noise control, use of hazardous materials, etc; welfare and hygiene facilities – drinking water, toilets, washing facilities, food preparation, canteen/mess room, first aid, personal protective clothing, etc; site organisation – good order, tidiness, sign-posting and lines of demarcation.

Storage and protection of materials

In Chapter 1 some aspects of direct waste were considered, with reference to bad handling and storage. Here, the more common construction materials are reviewed with regard to recommended storage procedure.

Cement

Construction of large reinforced concrete buildings will justify bulk storage of cement in cylindrical containers, known as silos, above the concrete mixing plant. Cement in silos is well protected and replenished on a daily basis, but where smaller batches of concrete or mortar are required, cement is supplied in bags. Cement will absorb moisture readily, therefore storage time should be minimal, otherwise mortar or concrete strength could be seriously impaired.

Dry storage for the 50 kg bags is preferably in a shed or surplus site hut with a raised floor. Bags should be kept clear of the shed side walling as dampness easily penetrates the thin timber cladding. In the absence of a shed, the minimum effective cover is a base of timber bearers and boarding to support the cement clear of ground dampness, and a tarpaulin or plastic sheet covering tied and staked to the ground as shown in Fig. 2.2.

Cement should be used in delivery date rotation, to avoid long storage and possible interference with the setting chemistry.

Aggregates

Aggregates are the stones and sands used for concrete and mortar production. They are not likely to be damaged by moisture, but water trapped between the voids will seriously affect concrete strength if allowance is not made during mixing. The water/cement ratio is critical for structural concrete, therefore a sample of wet aggregate should be weighed and compared with the dry weight in order to modify the added water content. Covering aggregates is

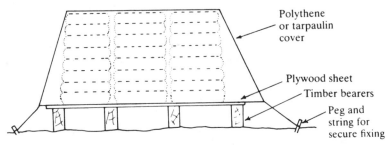

Fig. 2.2 External cement storage

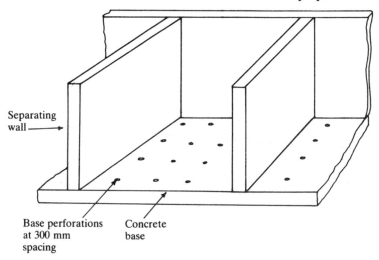

Fig. 2.3 Segregated aggregate store

not usually convenient but a clean hard base, preferably with drainage perforations will prevent mud and leaves contaminating the aggregates. Figure 2.3 shows a possible aggregate store with separating walls of timber or blockwork. Walls are essential to prevent intermixing of different aggregate grades.

Timber

After seasoning, timber has a moisture content of between 15 and 20%, a considerable reduction from its natural state. Therefore, water is readily absorbed, but this must be avoided as it affects the material's workability, causes deformities and may encourage rot. Site storage should be in a covered well-ventilated area, but if this is not convenient external storage as shown in Fig. 2.4 is acceptable. Gaps between adjacent sections encourage air circulation, and a waterproof cover is essential in case of rainfall.

Bricks and blocks

Bricks and blocks can be delivered packaged, with steel tape binding or on pallets. They should remain this way for easy site handling by crane or fork lifting plant. Stacking beyond head height is dangerous, particularly if the units are not bound and stacked loose. Covering with plastic sheeting is necessary to prevent saturation. This creates difficulties in laying, particularly with lightweight porous blocks which could also suffer frost

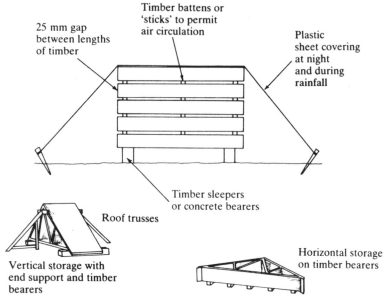

Fig. 2.4 Storage of timber

damage. A dense stable base is also necessary for safety and to prevent mud contamination.

Plaster

This has similar properties to cement, except the absorption of only a small amount of moisture will ruin the setting process and strength. Plaster must be stored internally and used within a few days of delivery, preferably the same day. Some manufacturers date-stamp the bags to provide guidance on suitability.

Plasterboard

Plasterboard is supplied in large sheets having limited flexibility. It must be handled with care, with particular regard to corners which are easily damaged. External storage is inadvisable as contact with water will decompose the plaster core. Plasterboard is a finishing material and can normally be taken directly into the building. Stacking is preferably flat, but if this is inconvenient, edge stacking is acceptable if not overloaded.

Steelwork (sections and reinforcement)

This is most conveniently stored in racks produced from scaffold tubing. Similar sizes and lengths can be placed together for easy identification. Ground storage is satisfactory, if components are raised from the mud by timber bearers or concrete blocks. Covering is unnecessary except for long storage periods, as slight surface rust increases the bond with surrounding concrete. The exception is steel sections to be painted, which should be kept dry unless pre-primed before delivery.

Scaffolding too, can be stored in a rack, with couplings and similar fittings kept together in boxes.

Trial boreholes

Trial holes to determine the nature of a subsoil are an important part of an early site investigation. More detailed and thorough examination will occur later when information about building design and structural loading can be related to the sub-soil's bearing potential. Preliminary examinations may be with trial pits excavated by spade or a hand auger of the type shown in Fig. 2.5. Later, when more detailed information is required, a powered auger is more effective. The pile boring apparatus shown in Fig. 3.13 could be employed for this purpose.

The depth of boreholes can be several metres for high-rise

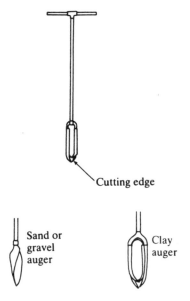

Cutting edge

Sand or gravel auger

Clay auger

Fig. 2.5 Hand augers for site investigation

Site High Street, Anytown. Type of boring Shell and auger. Ground level 53.65 m above ordnance datum. B.H. Diameter 200 mm. Remarks Borehole lined.				Borehole N°: & Date of boring 10.10.86 Water level 52.15 m		

Description of strata	Sample	Depth	Legend	Depth (m)	Total (m)
Topsoil					
Composite of clay and rubble	D	1.000		1.100	0.900
		1.500		WATER	TABLE
	U	2.000			
Black silt with gravel	U	3.000		3.140	2.040
Coarse silt				3.190	0.050
	D	4.000			
	D	5.000			
	U	6.000			
Soft black silty clay	U	7.000		7.140	3.950

Fig. 2.6 Borehole log

buildings; for small factories and houses the relevant depth is contained within the foundation pressure bulb diagram shown in Fig. 3.1. Borings can be at random or regular intervals. If the building position is known they should coincide with foundation positions, particularly wall junctions and intersections. Samples of subsoil can be extracted loose or disturbed, or undisturbed in steel tubes. They are recorded on a borehole log, shown in Fig. 2.6 and samples are removed for laboratory analysis to establish bearing capacity and chemical composition. Each borehole has a reference number to coincide with a position on the site plan. Figure 2.7 is included to show the recommended convention for representation of subsoil types.

Site fencing and hoardings

Provision of site fences or hoardings is controlled by the Highways Act, under the jurisdiction of the local authority highways department.

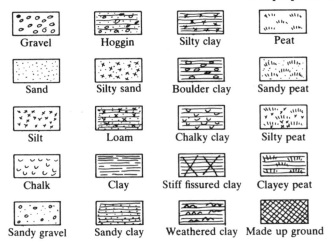

Fig. 2.7 Symbols for subsoils

Security of each site is considered on its merits, and usually the builder's concern for his stock and equipment results in a fencing system which more than satisfies the local authority's interests in public safety.

Where a hoarding is adjacent to or encroaching onto the footpath or highway the local authority may insist on a financial deposit, to cover the cost if the builder fails to reinstate any damaged surface.

The local authority can also insist that a temporary footway is provided outside the hoarding, having a handrail and suitable means of cover for pedestrian protection. It will also be necessary for the builder to maintain the hoarding in good and safe order, possibly with the provision of illumination under the cover and red hazard lighting around the perimeter. The builder is responsible for removal of the hoarding on completion of work. A fine is applicable for non-conformity with this section of the Highways Act, plus a further fine for every day the offence continues. If the hoarding is found unstable and unsafe, additional fines can be imposed.

The type and nature of fencing selected will depend considerably on the work composition and site location. Small housing work in rural areas rarely attracts any fencing legislation from the local authority, but the builder would be wise to use a simple open fence of the chestnut paling variety (see Fig. 13.1) to deter children.

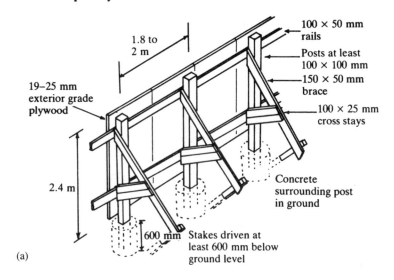

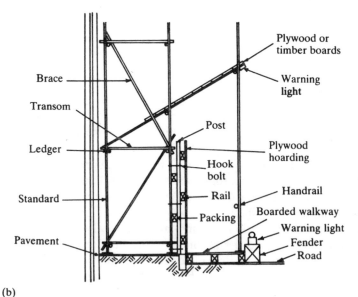

Fig. 2.8 Hoardings (a) free standing, (b) scaffold frame

Valuable materials can be locked in site accommodation or completed garages. In urban areas the local authorities will be more precise, requiring close-boarded hoarding of the type shown in Fig. 2.8 and its position clearly indicated on the site plan with coloured ink.

Close-boarded hoarding

This is far more effective than open fencing for protecting the site from intruders. The original construction was a series of boards close-butted and nailed to a framework, hence the term close-boarded, but now plywood sheet of 2.4 m height by 1.2 m width is ideal for preventing viewing through gaps or over the top. Sheets are usually nailed to posts or rails, but may be attached to a scaffold frame where this encroaches onto the pavement. In this latter case a fan guard or fan hoarding inclines over the temporary pedestrian access to protect people from falling materials. A fan variation is shown in Fig. 2.9, using demolition materials retained to the existing lower floor and wall.

Excavation and site clearance

Excavation and site clearance can involve several different items of plant. Most plant manufacturers have responded to this difficulty by rationalising their design on the basis of a standard agricultural tractor. Changeable attachments at the front and back achieve sufficient versatility and adaptability to fulfill all but the most difficult of site tasks. Large-scale construction and motorway development is excepted, but for the benefit of this section, site

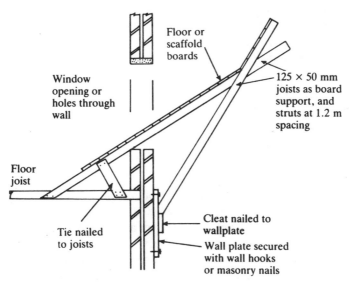

Fig. 2.9 Fan hoarding constructed from demolition materials

preparation for housing and light industrial development is assumed.

Types of excavation

The excavation technique is largely determined by plant availability and subsoil composition. Convenient classification includes:

1 Hand tools
2 Mechanical plant

Hand tools

Excavation using spades, hand augers, picks and other manual implements is virtually obsolete. The exceptions are:

(a) very small buildings, e.g. garage, or house extension;
(b) where the site is inaccessible to excavating plant;
(c) where archeological remains are discovered and particular care is necessary.

Otherwise the only use of hand tools is for trimming excavations mechanically where awkward projections and deviations are specified.

Mechanical plant

Mechanical plant and equipment save considerable man-hours, and are standard features on all sites. The type of plant varies with the nature of work and the different construction stages. Plant is most commonly used for:

(a) site clearance and light demolition;
(b) topsoil stripping;
(c) trench excavation;
(d) pits and boreholes;
(e) site transport.

Site clearance

Removal of hedges, trees, existing buildings and undulations is the first of site operations, to achieve a clear uninterrupted work space. Most tractors with face-shovel attachments will be capable of pushing out trees and shrubbery. A back-hoe attachment is useful for digging out stumps and roots. Chains or wires secured to a tractor for pulling out trees is also successful for moderately-sized growths.

Burning is acceptable if adequate control can be achieved, particularly for tree stumps where chemical decomposition is considered too slow. Buildings can be demolished slowly by hand or by rapid destruction. Hand demolition is usually preferred as most building materials have a good second-hand value and the process is easy to control. Destruction or fragmentation by fire or explosion is rapid, but imposes some difficulty in disposing of the rubble. A cast-iron or concrete ball swinging from the jib of a crane is more methodical, with a face shovel and lorries to remove the surplus materials. Small buildings can be levelled with a dozer or tractor with dozing attachment.

Levelling, land clearance and stripping of the topsoil are all easily achieved with a bulldozer. The bucket or blades vary to suit soil composition and are usually dished to encourage the soil to roll off rapidly. With the blade lowered and angled as shown in Fig. 2.10, slopes may be cut, and the surplus spoil used to fill or level. This is particularly useful for terracing sloping sites.

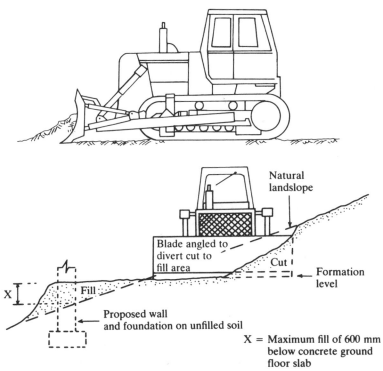

Fig. 2.10 Bulldozer for site levelling

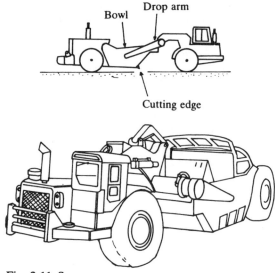

Fig. 2.11 Scraper

Very large sites, airfields or motorways are more effectively levelled with motorised scrapers. An example is shown in Fig. 2.11 containing a large bowl with lowered cutting edge for obtaining precise formation levels. As the bowl fills, a hydraulic ram displaces the spoil.

Stripping topsoil

Removal of topsoil or vegetable soil is essential before construction work commences. It is soft, easily compressed and contains plant life, which are all unacceptable properties as the basis of construction. It is stripped, usually to a depth of about 200 mm, and retained for reuse when the site is landscaped on completion. Plant previously described for site clearance, i.e. dozer and face shovel (Fig. 2.13), are ideal for small sites.

Trench excavation

Most trench excavation for services and foundations is with a back hoe or backacter. Small machines are wheeled tractors, and larger versions are tracked. They span the proposed trench and the toothed bucket and hydraulic boom extend out and excavate towards the cab. Most spoil is retained for backfilling. They are also adaptable for basements and ditches, and Fig. 2.12 shows typical range dimensions for an average tractor-based machine.

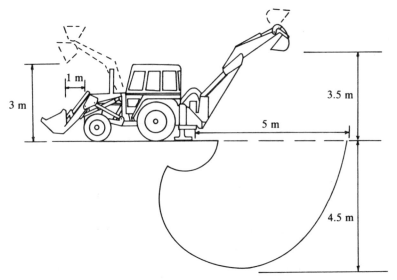

Fig. 2.12 Excavator/loader (typical operating dimensions)

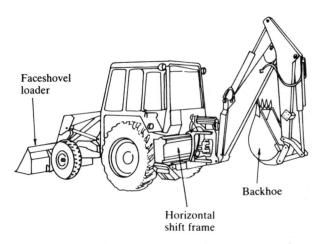

Faceshovel loader

Backhoe

Horizontal shift frame

Fig. 2.13 Loader/backhoe excavator

Figure 2.13 is more illustrative, indicating the interchangeable (varying widths and capacities) bucket and the front shovel for topsoil stripping, light demolition and loading functions.

Pits and boreholes

Large pit excavation is possible with back-acting equipment shown

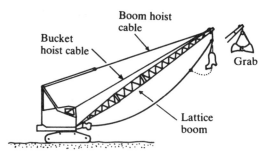

Fig. 2.14 Dragline/grabcrane

in Figs 2.12 and 2.13. However, where the volume is large the crane-mounted dragline shown in Fig. 2.14 is preferred. The bucket is swung forward to penetrate the subsoil and dragged back towards the cab. Discharge of spoil is to one side by forward tilting of the bucket. Deep excavation into granular soils is more effective with a grab or 'clamshell' operating by gravitational penetration.

Boreholes for piled foundations and subsoil samples are by hand auger (Fig. 2.5) or powered auger (Fig. 3.13). An alternative in soft clayey subsoils is a tripod-suspended shell auger shown in Fig. 2.15. The steel shell is raised by a power winch and dropped by gravity. Subsoil is retained inside the shell by a non-returning clack valve, and removed through a side access door. These borings require a steel lining to prevent the void collapsing.

Site transport

For transportation of most materials and light equipment, the site dumper shown in Fig. 2.16 is ideal. It has been the site 'workhorse' for a considerable period, and modern improvements enhance its adaptability. These include two- or four-wheel drive variations,

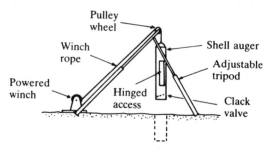

Fig. 2.15 Shell auger rig

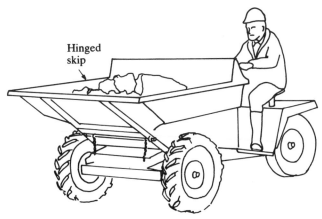

Hinged skip

Fig. 2.16 Site dumper

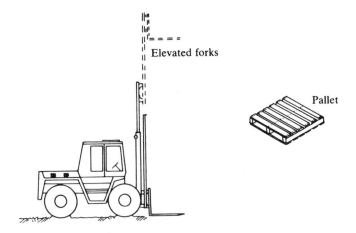

Elevated forks

Pallet

Fig. 2.17 Fork-lift material transport

with skip (rotational if required) capacities from 0.3 to 3 m^3. The dumper's versatility has been challenged in recent years by the site fork lift machine shown in Fig. 2.17. These are similar to the factory versions, but have larger wheels. Fork lifting is practical because many materials are delivered on pallets, and the lifting facility can be used for direct loading onto a scaffold frame.

Site haulage lorries are seen at an early stage of site preparation for removing surplus spoil and demolition materials. Capacities range from 5 to about 20 tonnes and they all feature a tipping body as shown in the examples in Fig. 2.18.

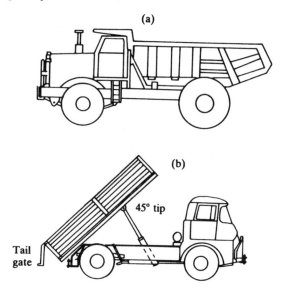

Fig. 2.18 Haulage vehicles (a) Tipping, dumper truck
(b) Tipper/transporter truck

Levelling

Levelling is the procedure for measuring the height of certain points around the site, in order to establish excavation quantities for reducing ground levels and establishing drain runs. The height is measured in relation to ordnance bench marks, shown in Fig. 2.19. These are cut into vertical brick or stonework at approximately 300 m intervals in towns, and 1000 m intervals in rural areas. By reference to the appropriate ordnance survey map the height above mean sea level at Newlyn may be established. If the site is not conveniently located to an ordnance bench mark, a temporary bench mark can be created from a manhole cover or other site fixture to which all measurements relate.

Measurement of heights above the reference point is undertaken with a builders' level and a staff. The level is basically a telescope mounted on a tripod, with cross-hair lines to establish focus and staff readings. The staff is produced in 3, 4 or 5 m lengths with folding or telescopic facilities for easy transportation. It is graduated in 10 mm intervals as shown in Fig. 2.19. Staff readings are taken at pegged points around the site to correspond with a grid or other convenient arrangement shown on the site

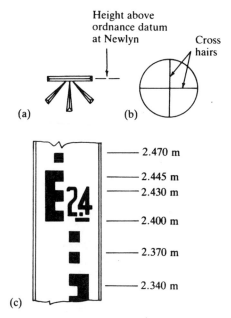

Fig. 2.19 Site levelling equipment (a) Bench mark (b) Telescopic
markings as viewed through level (c) Staff markings
and example readings

plan. By establishing a regular pattern of readings it is possible to
calculate excavation volumes. Figure 2.20 shows the principle of
taking staff readings and determining the difference in levels from
the initial reference.

Setting out

Following stripping of the topsoil and levelling, the position of a
building is marked out with string lines and pegs to indicate
foundation trenches and walls. A frontage line is obtained from
the site plan. This indicates the position of the building relative to
a kerb line or centre line of the road as shown in Fig. 2.21. Also
shown is the building line; this is the closest position that the local
authority will permit construction relative to the kerb or road
centre line.

Having established the frontage line with stout pegs and string,
perpendicular ranging lines locate the front corners. The rear of
the building is similarly positioned. Accuracy should be achieved

(a)

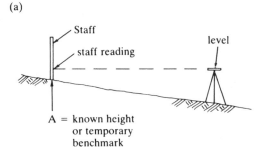

(b)

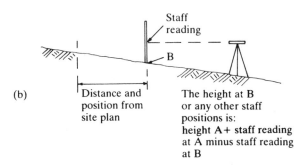

Fig. 2.20 Site levelling principles (a) First reading (b) Second and subsequent readings

by nail positioning the string in pegs, and squareness of corners by a large timber square or use of Pythagoras' theorem to obtain a right-angled triangle having sides in the ratios of 3:4:5. Figure 2.22 illustrates these principles with diagonal measurements across the corners as an additional check for squareness. Offsets are added and checked in the same manner, and projections or bays (particularly if angular) are positioned with a temporary timber frame work to follow the bay profile.

When the building outline is finally checked and accurately positioned, profile boards are erected clear of the string lines. These are horizontal boards nailed to strong pegs with nails or saw cuts in the upper edge to receive the original string lines. Figure 2.23 shows the arrangement of boards at corners, where they are also used to mark the position of trench lines. Trenches are then represented with cement or sand placed along the ground surface beneath the string line to provide guidance for the excavator when the string is removed. After excavation and concreting for

(a)

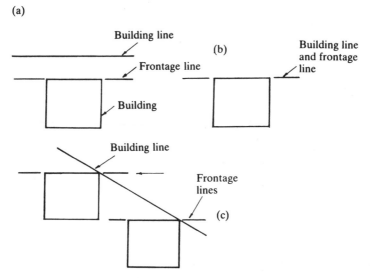

Fig. 2.21 Location of frontage lines (a) Frontage line set behind building line (a) Building line and frontage line coincide (c) Frontage line angled to building line

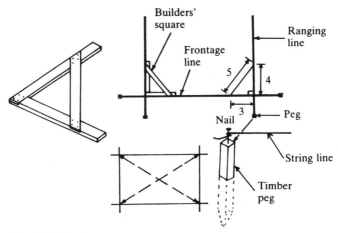

Fig. 2.22 Setting out corners

foundations, the profile boards are re-established and string lines suspended to locate the inner and outer positions of walls. To commence bricklaying, the string positions are transferred to the concrete foundation by use of a plumb bob or spirit level as shown in Fig. 2.23. Here a thin skim of mortar is scored to represent the

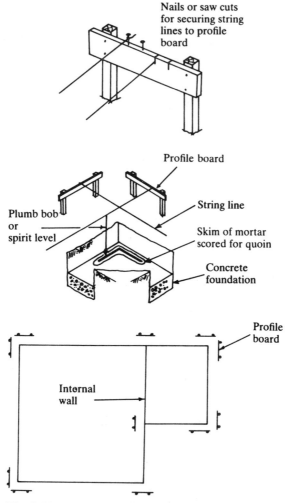

Fig. 2.23 Location of profile boards

corner. Once the brickwork is well established, string lines are suspended from each corner to ensure brick alignment and the profile boards are no longer required.

Safety in excavations

Excavations for drain and foundation trenches are one of the most dangerous areas where site operatives are employed. The safety legislation is contained in the Construction (Health, Safety & Welfare) Regulations, now controlled by the Health and Safety at Work, etc. Act. This emphasises the need for assessment of risk to personnel in open trenches, which is likely to require an adequate supply of timber supports and props to withstand earth pressures. Exceptions may be:

1. If the trench is shallow and the sides are self supporting.
2. Where the sides of the excavation are sloped below the subsoil angle of repose, (self supporting angle).
   Examples of angles of repose; dry clay – 46°
   wet clay – 16°
   dry sand – 40°
   wet sand – 22°

Both exceptions are illustrated in Fig. 2.24. Further good practice includes:

1. Daily inspection of excavations by a competent person to ensure safe working conditions.
2. A fence or barrier around all excavations.
3. No materials deposited within 1.5 m of the trench.

These requirements are also shown in Fig. 2.24.

Support to shallow excavations

Trench support has acquired the term 'planking and strutting', implying timber components. Whilst timber is often the most convenient material for shallow trenches, steel interlocking polings are often used for deep water-logged subsoils. Adjustable steel struts are also more convenient and have considerable reuse value for all depths of excavation; these are shown in Fig. 2.25 with corresponding size and span.

Other examples of trench timbering shown are more traditional. These contain polings of 175 to 225 mm width by 35 to 50 mm thickness in lengths up to about 1.5 m to suit the trench depth. In reasonably firm subsoil, polings face the subsoil with direct support from struts or longitudinal support from horizontal walings. Used in this fashion, walings range in size from 75 × 100 mm to 100 × 175 mm. In loose or saturated subsoils, walings may be used to totally sheet or board the trench by directly facing the subsoil. They can be installed in stages with temporary struts as the trench is deepened and in this sheeting role could range between 38 and 75 mm thick × 225 mm wide. Thickness will be

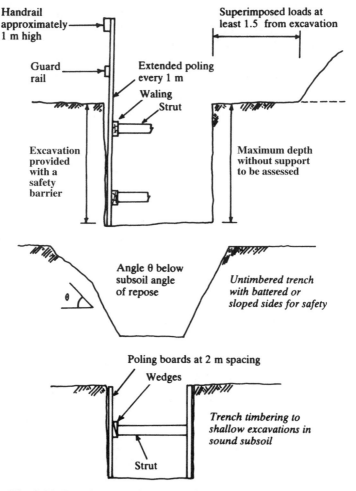

Fig. 2.24 Trench excavations and safety features

determined by spacing of struts which should not exceed 2 m. Spacing much less than 2 m is also undesirable as this adds congestion to a very restricted work space. Struts are designed to restrict opposing subsoil thrust and earth pressure. If the trench is cut to a slight taper, struts may be over-cut in length and driven in place by sledge hammer. Otherwise, they are cut slightly under length and wedged between waling or poling as shown in Fig. 2.25. Square timber of 75 × 75 mm or 100 × 100 mm section is usually most suitable.

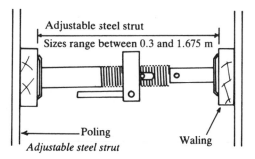

Adjustable steel strut

Sizes range between 0.3 and 1.675 m

Poling

Waling

Adjustable steel strut

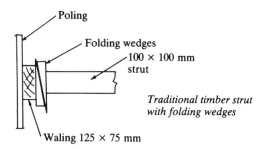

Poling

Folding wedges

100 × 100 mm
strut

*Traditional timber strut
with folding wedges*

Waling 125 × 75 mm

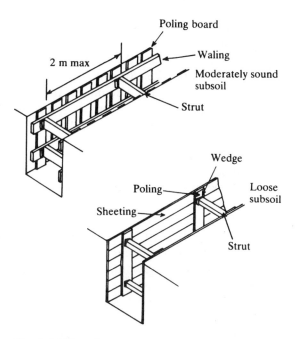

Poling board

Waling

2 m max

Moderately sound
subsoil

Strut

Wedge

Loose
subsoil

Poling

Sheeting

Strut

Fig. 2.25 Trench support systems

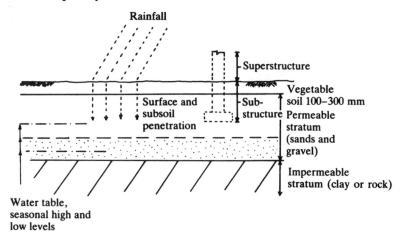

Fig. 2.26 Substructure and water table

Ground-water control

Excavation or sample boreholes frequently locate a level of saturation within a few metres of the surface. This is known as the water table, and is represented in seasonally variable positions (high after winter, low after summer), in Fig. 2.26.

Excavation and subsoil control below the water table is difficult and inconvenient. Furthermore, the strength of any concrete placed in water will be seriously impaired. Removal of ground-water may be temporary, i.e. for the duration of substructural working only, or permanent.

Temporary ground-water control

This is the simpler method; it involves digging a small sump within or close to the general excavation, to a depth in excess of substructural formation level. This arrangement is shown in Fig. 2.27 with a diesel powered diaphragm pump at surface level and a strainer or filter in the sump attached to the pump suction pipe. Electric centrifugal pumps are more effective for clear water, but this condition is unlikely in muddy excavations, and electricity may not be available on site at this stage. Nevertheless, where electricity is available, the use of specially-designed submersible pumps with wide gaps between the impeller blades for coping with muddy water may be preferred, and this is also shown in Fig. 2.27.

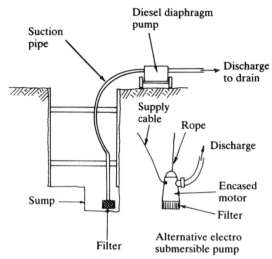

Fig. 2.27 Water control in excavations

Permanent ground-water control

If the level of ground saturation is noticeably high, it must be permanently lowered to satisfy Part C of the Building Regulations. This is to reduce substructural water pressure and the potential damaging effect of frost expansion or heave.

The oldest and simplest method of ground-water control is provision of ditches laid to a fall to carry ground-water away. However, these are potentially hazardous and soon become overgrown if not maintained. An improvement is a French drain shown in Fig. 2.28. These are basically ditches excavated to a gradient and filled with rubble. The illustration includes three grades of granular fill to function as a surface water cut off, a filter and a carrier. Topsoil may replace the granular fill to the surface to preserve uniform appearance, but this will reduce the effectiveness for collecting surface water. The life of these drains is limited as they become clogged with fine particles of silt, therefore installation with a permanent perforated or porous pipe void is preferred.

Land drainage pipes

Land drains offer a permanent drainage channel, eventually discharging into a stream or other convenient outfall. Pipe materials include porous or perforated clayware shown in

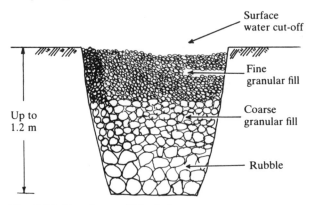

Fig. 2.28 French or rubble drain

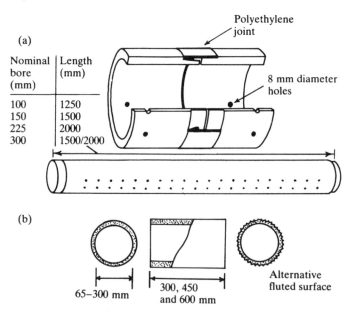

Fig. 2.29 Clayware subsoil drain pipes (a) Perforated (b) Porous

Fig. 2.29, porous concrete (no-fine aggregate), perforated pitch fibre and perforated uPVC all shown in Fig. 2.30.

The bedding method depends on whether the pipe is to collect and transmit ground-water only, or whether it is to function with a granular drain trench and carry additional surface water. Both methods are shown in Fig. 2.31. On long drain runs a silt catch pit is necessary to collect and trap fine materials to preserve the drain

(a) Porous concrete

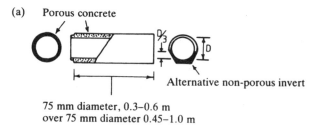

Alternative non-porous invert

75 mm diameter, 0.3–0.6 m
over 75 mm diameter 0.45–1.0 m

(b) 75,100 and 150 mm

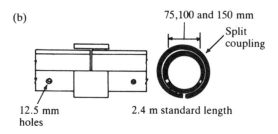

Split coupling

12.5 mm holes

2.4 m standard length

(c)

Slot length 25 mm min. for 100 mm pipe,
43 mm min. for 150 mm pipe.

Spigot and socket joint

75 mm   Slot width

3– 4 mm

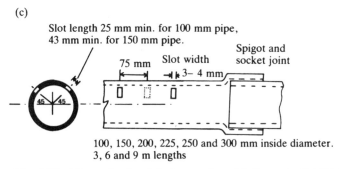

100, 150, 200, 225, 250 and 300 mm inside diameter.
3, 6 and 9 m lengths

Fig. 2.30 Alternative land drainage pipes (a) Concrete (b) Pitch
fibre (c) uPVC

efficiency. A simple traditional brick and concrete pit is shown in
Fig. 2.32 with sufficient access space for cleaning at periodic
intervals.

Systems

The layout of land drains will be determined by the land formation
and the location of buildings. Some formal systems are shown in
Fig. 2.33. The herringbone, parallel and fan arrangements suit a
sloping site where tributary or branch drains feed one main drain.
The grid pattern is the most regular, having the main drain to one

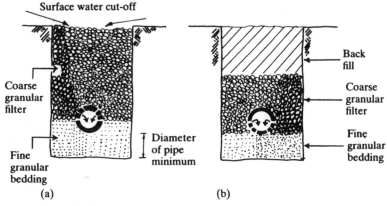

Surface water cut-off

Coarse granular filter

Fine granular bedding

Diameter of pipe minimum

Back fill

Coarse granular filter

Fine granular bedding

(a)   (b)

Fig. 2.31 Perforated pipes as surface and subsoil drains (a) Surface water and ground water drain (b) Ground water drain

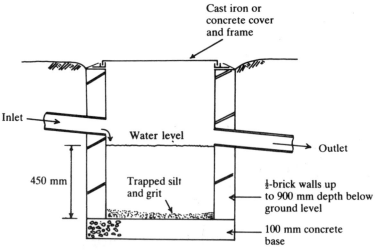

Cast iron or concrete cover and frame

Inlet

Water level

Outlet

450 mm

Trapped silt and grit

½-brick walls up to 900 mm depth below ground level

100 mm concrete base

Fig. 2.32 Silt trap or catch pit

side of the site and branches at right angles between buildings. The moat system is a complete perimeter drain around a building, ideal for intersecting ground-water to basements, and often installed as part of a renovation programme. A semi-moat system is acceptable if ground-water is flowing in one direction.

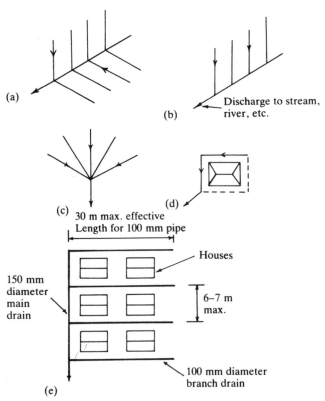

Fig. 2.33 Land drainage systems (a) Herringbone (b) Parallel
(c) Fan (d) Moat (e) Grid

3

Substructure

Factors affecting foundation design

The foundation is an integral part of a building that transfers the structural load to the ground. Selection of foundation type and design depends on two distinct variables:

1   The total building load.
2   The nature and quality of the subsoil.

It is essential to achieve a satisfactory balance between these two conditions, otherwise overstressing of the subsoil will lead to excessive building settlement and possible serious structural defects. A foundation must therefore safely sustain the dead, imposed and wind loads and transmit these to the ground without impairing a building's stability.

The total building load

The total load that a building transfers to the subsoil is composed of dead load, imposed load and wind load. Dead load is the force attributed to the structural mass of the construction. This refers to the combined weight of bricks, concrete, timber, tiles, etc. converted to units of force. (Note: $kg \times 9.81 = N$.) Imposed load results from the combined force of non-permanent fixtures and fittings such as furniture and persons using the building. For design purposes $1.5 \ kN/m^2$ is permissible. Snow load is also an imposed loading, and an allowance of $1.5 \ kN/m^2$ for flat roofs and $0.75 \ kN/m^2$ for pitched roofs over $30°$ should be included.

Wind loading is difficult to define as there are numerous variables to combine before a factor relating to a dynamic force can be achieved. The procedure is explained in BS CP3 : Ch. V : Pt.2 : 1972 (*Wind Loading*), using maximum wind speed as the basis of calculation with multipliers for topography, ground roughness, exposure period and pressure coefficients. For low-rise dwellings of regular shape in moderately sheltered positions, the effect is minimal and unlikely to exceed 1 kN force per square metre of surface.

The nature and quality of the subsoil

Before deciding on which type of foundation to use, it is essential to examine the subsoil material below the level of the proposed foundation. Normally, strip foundations are unlikely to extend beyond 1 m depth, therefore the area immediately below this will be subject to most pressure, gradually dispersing within the bulb of distribution shown in Fig. 3.1. Also shown is the area of greatest shear, a semicircle of diameter equal to the foundation width.

Subsoil samples are located by hand or power auger and extracted by sampling tubes for laboratory analysis, or subject to *in situ* field tests to ascertain their bearing capacity. Table 3.1 shows simple field tests and provides a comparison between subsoil materials, total load per metre run of foundation and foundation width. Alternatively foundation width may be obtained by dividing the load per metre run by the analysed safe bearing capacity of the subsoil as calculated in Fig. 3.2.

Subsoil movement

To understand the factors affecting subsoil movement it is essential to appreciate the difference between cohesive and non-cohesive subsoil materials. Cohesive subsoils contain very fine

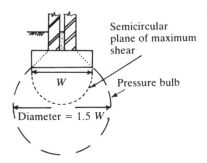

Fig. 3.1 Zones of subsoil affected by pressure and shear transfer from a strip foundation

Table 3.1 Strip foundation width relative to subsoil quality

Type of subsoil	Condition of subsoil	Field test applicable	Minimum width in millimetres for total load in kN/m of loadbearing walling of not more than						
			20	30	40	50	60	70	
I Rock	Not inferior to sandstone, limestone or firm chalk	Requires at least a pneumatic or other mechanically operated pick for excavation	In each case equal to the width of wall						
II Gravel Sand	Compact Compact	Requires pick for excavation. Wooden peg 50 mm square in cross-section hard to drive beyond 150 mm	250	300	400	500	600	650	
III Clay Sandy clay	Stiff Stiff	Cannot be moulded with the fingers and requires a pick or pneumatic or other mechanically operated spade for its removal	250	300	400	500	600	650	

IV Clay Sandy clay	Firm Firm	Can be moulded by substantial pressure with the fingers and can be excavated with graft or spade	300	350	450	600	750	850	
V Sand Silty sand Clayey sand	Loose Loose Loose	Can be excavated with a spade. Wooden peg 50 mm square in cross-section can be easily driven	400	600					Note: Foundations do not fall within the provisions of regulations if the total load exceeds 30 kN/m
VI Silt Clay Sandy clay Silty clay	Soft Soft Soft Soft	Fairly easily moulded in the fingers and readily excavated	450	650					Note: In relation to types VI and VII, foundations do not fall within the provisions of regulations if the total load exceeds 30 kN/m
VII Silt Clay Sandy clay Silty clay	Very soft Very soft Very soft Very soft	Natural sample in winter conditions exudes between fingers when squeezed in fist	600	850					

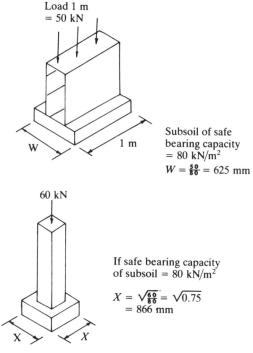

Load 1 m
= 50 kN

Subsoil of safe
bearing capacity
= 80 kN/m^2
$W = \frac{50}{80} = 625$ mm

60 kN

If safe bearing capacity
of subsoil = 80 kN/m^2

$X = \sqrt{\frac{60}{80}} = \sqrt{0.75}$
$= 866$ mm

Fig. 3.2 Foundation design calculations

particles, relying on water for strength and bonding. They are predominantly clay but may contain sand and other larger particles which will improve their bearing capacity. These subsoils have high swelling and shrinkage characteristics and are particularly responsive to dehydration from drought or extraction of ground-water by trees. Also, ground swell could follow felling of large trees and is very likely to occur where new building on a sloping site provides a superimposed load as shown in Fig. 3.3. Construction on cohesive subsoils should never be closer than the mature height of an adjacent tree and 1½ times the height of a group of trees. Physical damage to foundations by tree roots is also possible, particularly from willow, elm and poplar. To avoid damage, foundations should be reinforced by a deep strip or beam of at least 900 mm depth and about 400 mm width as shown in Fig. 3.10. Alternatively, foundations could be piled down to

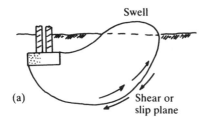

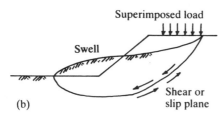

Fig. 3.3 Shear failure of subsoils (a) Cartwheel failure (b) Failure on inclined sites

load-bearing strata and provided with an edge beam of the type shown in Fig. 3.14.

Non-cohesive sandy subsoils are less of a problem. The particles are unaffected by water and rely on size and closeness of compaction for strength. Consolidation due to building load is rapid, therefore settlement cracking of the superstructure should be insignificant. Sloping sandy sites can prove difficult due to running ground-water which may erode the finer particles, leaving the coarser material in an unstable condition. Also, subsoil drainage will be required where the water table is high, otherwise winter freezing will increase the ground volume and lift the surface. To avoid damage by frost heave a foundation with a depth of at least 600 mm is necessary. The traditional strip foundation shown in Fig. 3.6 is normally quite adequate.

Foundation types

Foundations for low-rise construction are principally:

1   Strip: (a) traditional shallow;
           (b) wide;
           (c) deep.

2   Raft.
3   Short-bored pile: (a) cast *in situ*;
    (b) precast.
4   Pad.

Strip foundations (see Fig. 3.4)

This type of foundation is a continuous level support for load-bearing walls. It is produced from concrete of minimum strength 15 N/mm^2 (1 : 3 : 6), and may be reinforced for poor subsoils or high loading. Reinforced concrete should have a strength specification of at least 25 N/mm^2 (1 : 2 : 4) which is less flexible, but the higher cement-to-aggregate ratio protects the steel from corrosion. Before specifying concrete for foundation work it is essential to determine the chemical composition of the subsoil. Concrete is resistant to most forms of chemical attack, but noticeably weak in the presence of sulphates. These combine with the aluminate content of cement, expand, and generate cracks and splits in the concrete. Concrete in these subsoils must be produced from sulphate-resisting Portland cement and not ordinary Portland cement.

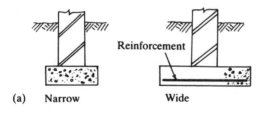

(a)   Narrow   Wide

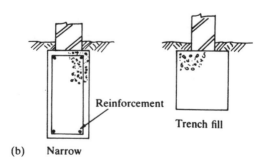

(b)   Narrow

Fig. 3.4 Types of strip foundation (a) Shallow (b) Deep

Traditional shallow strip

A traditional strip foundation is produced by excavating a trench of sufficient depth to avoid foundation damage from subsoil swelling and shrinkage, and of adequate width to spread the structural load over the required bearing area of subsoil. The bottom of the trench is compacted and at least 150 mm of concrete is placed and levelled as shown in Fig. 3.5. Figure 3.6 illustrates the completed substructure with a typical shallow strip foundation and Figs 3.7 and 3.8 the variations for sloping sites and requirements for attached piers respectively.

Wide strip

This type of foundation is used where the structural loading is very high relative to the subsoil-bearing capacity. Figure 3.9 shows the two possible forms of construction, unreinforced with the thickness at least equivalent to the projection, and reinforced in the area of greatest tensile stress to economise in concrete.

Deep strip

Deep strip foundations are a relatively new concept and have two applications:

(a) trench fill;
(b) reinforced.

(a) Trench fill is a very simple form of concrete foundation, designed to save considerable substructural construction time.

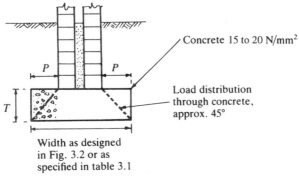

Concrete 15 to 20 N/mm^2

Load distribution through concrete, approx. 45°

P P

T

Width as designed in Fig. 3.2 or as specified in table 3.1

T, the thickness at least 150 mm or P, the projection. Take greater value.

Fig. 3.5 Building regulations affecting foundation dimensions

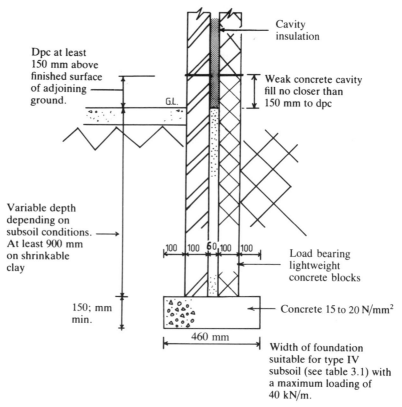

Dpc at least
150 mm above
finished surface
of adjoining
ground.

Cavity
insulation

Weak concrete cavity
fill no closer than
150 mm to dpc

G.L.

Variable depth
depending on
subsoil conditions.
At least 900 mm
on shrinkable
clay

100 100 60 100 100

Load bearing
lightweight
concrete blocks

150; mm
min.

Concrete 15 to 20 N/mm²

460 mm

Width of foundation
suitable for type IV
subsoil (see table 3.1) with
a maximum loading of
40 kN/m.

Fig. 3.6 Narrow strip foundation

Following trench excavation to the required design width and depth, concrete is placed to within two brick courses of finished ground level. Concrete costs are high, but savings in bricks, bricklaying time and trench timbering more than compensate. Furthermore the inconvenience and delays associated with bad weather at this stage are also avoided.

(b) Reinforced deep strip foundations are an acceptable alternative to wide strip foundations for soft clay subsoil conditions. The depth should be at least 900 mm to avoid the effects of shrinkage and swelling and about 400 mm wide to provide sufficient support for the wall. A narrow strip is adequate, as resistance to settlement is achieved by frictional resistance between the sides of the foundation and surrounding subsoil and by longitudinal bearing. Reinforcement is required as subsoils

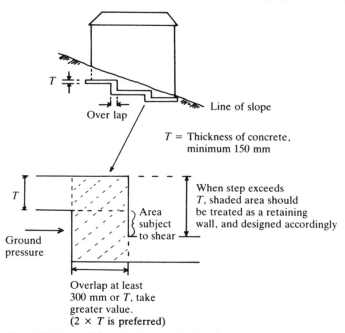

T = Thickness of concrete, minimum 150 mm

When step exceeds T, shaded area should be treated as a retaining wall, and designed accordingly

Overlap at least 300 mm or T, take greater value. ($2 \times T$ is preferred)

Fig. 3.7 Stepped foundation on sloping site

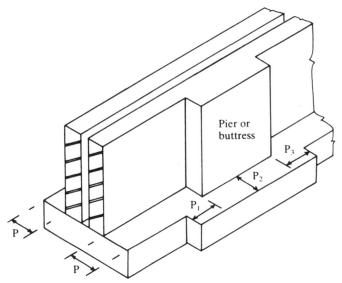

Fig. 3.8 Buttresses and attached piers. (Note: P = Projection of concrete beyond wall and, P_1, P_2 and P_3 are projections at least equal to P)

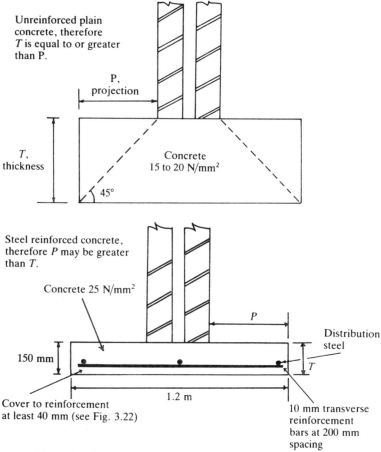

Unreinforced plain
concrete, therefore
T is equal to or greater
than P.

P,
projection

T,
thickness

Concrete
15 to 20 N/mm²

45°

Steel reinforced concrete,
therefore *P* may be greater
than *T*.

Concrete 25 N/mm²

P

Distribution
steel

150 mm

T

1.2 m

Cover to reinforcement
at least 40 mm (see Fig. 3.22)

10 mm transverse
reinforcement
bars at 200 mm
spacing

Fig. 3.9 Wide strip foundations

prone to volume change may develop voids in long periods of dry
weather. Figure 3.10 illustrates possible construction for use on
shrinkable clay subsoil.

Raft foundations

A raft foundation covers an area at least equal to the base
structural area of a building. In simplest form this is no more than
a 100 mm layer of concrete on hardcore, ideal as a temporary base
for site huts or as a permanent firm ground structure for garden
sheds and similar outbuildings. As a permanent ground location
for habitable accommodation this is unacceptable, as frost heave

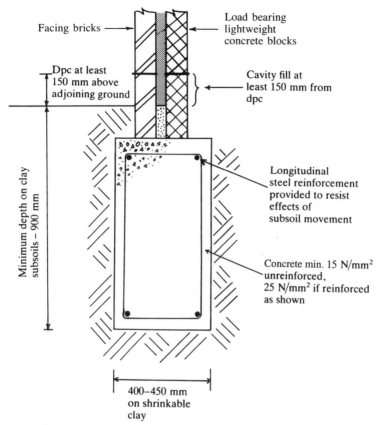

Facing bricks

Load bearing
lightweight
concrete blocks

Dpc at least
150 mm above
adjoining ground

Cavity fill at
least 150 mm from
dpc

Minimum depth on clay
subsoils – 900 mm

Longitudinal
steel reinforcement
provided to resist
effects of
subsoil movement

Concrete min. 15 N/mm²
unreinforced,
25 N/mm² if reinforced
as shown

400–450 mm
on shrinkable
clay

Fig. 3.10 Reinforced deep strip foundation

could lift the edge and cause structural failure. As foundations on
soft compressible subsoils concrete rafts have an application to
dwellings, providing they are well reinforced to resist the effect of
ground movement and are constructed with an edge apron to resist
sliding. Figure 3.11 shows a possible construction for use with a
single-storey structure on soft clay or peat subsoil.

Short-bored piled foundations

Piled foundations have an application when conventional
foundations would otherwise be very deep as it is uneconomical to
consider normal excavation beyond about 2 m. The type of subsoil
conditions will include shrinkable clays, infill or waste tips, slopes
and high water table in poorly drained areas. The system

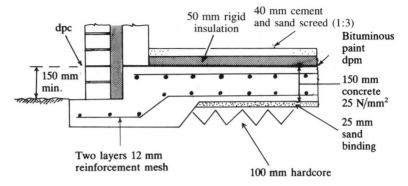

Fig. 3.11 Reinforced concrete raft foundation

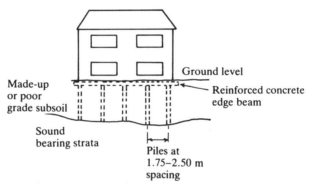

Fig. 3.12 Piled foundation

resembles building on stilts or columns, with the base of the support resting on a load bearing stratum up to about 4 m below the surface. Figure 3.12 shows the principle, which is the same for cast *in situ* or precast driven piles.

Cast in situ

Cast *in situ* piles are generally end bearing, but may have enlarged diameter for frictional resistance to settlement where quality subsoil is impossible to locate. Boring is by powered auger, as shown in Fig. 3.13 to a depth rarely exceeding 4 m. Beyond this the expense becomes prohibitive for construction of simple dwellings. A steel tube lining is frequently required where the boring is liable to collapse, and this should be extracted after reinforcement and concrete are placed. Figure 3.14 shows the

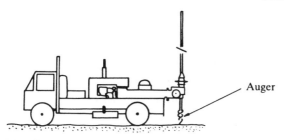

Fig. 3.13 Lorry-mounted auger

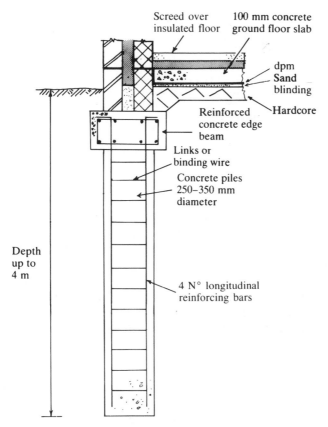

Fig. 3.14 Bored and reinforced concrete pile foundation cast *in situ*

construction of a typical cast *in situ* end bearing pile with longitudinal reinforcement to resist the effects of ground movement.

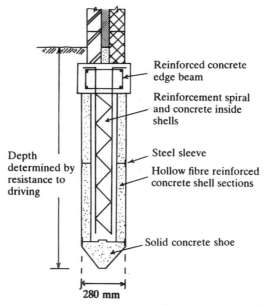

Reinforced concrete
edge beam

Reinforcement spiral
and concrete inside
shells

Steel sleeve

Hollow fibre reinforced
concrete shell sections

Depth
determined by
resistance to
driving

Solid concrete shoe

280 mm

Fig. 3.15 Precast driven **shell pile** foundation

Precast concrete piles

These are fibre-reinforced concrete hollow shells of approximately
300 mm diameter and 1 m length, linked with a steel sleeve.
Placing is by driving each section with a vertical drop hammer until
sufficient resistance is achieved. The hollow core receives
reinforcement and *in situ* concrete which ties in with an *in situ*
concrete edge beam as detailed in Fig. 3.15. The potential in poor
subsoils is considerable, as boring equipment is not required and
depth of ground penetrations is less critical.

Pad foundations

Pad foundations associated with reinforced concrete or steel
framed building are considered in more detail in Chapter 7. They
have an application to dwellings where isolated brick columns
are required in order to preserve access under a first-floor extension as shown in Fig. 3.16. Calculation of pad bearing area is
slightly different to strip foundations. These are compared in
Fig. 3.2.

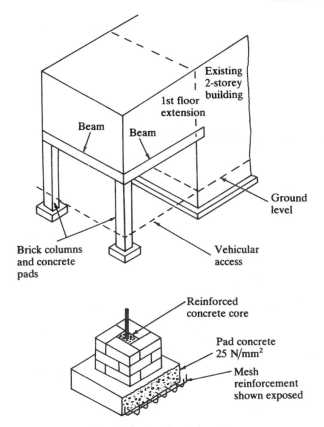

Fig. 3.16 Pad foundation and columns

Basement construction

Excavation

The method of excavation for basement construction will depend on the availability of space. The majority of structures requiring basements are likely to be in commercial centres with very restricted construction area. Here, excavation and construction will be close to, or at, the property boundary and the method employed will involve perimeter trench excavation. This is shown in principle in Fig. 3.17 with the alternative battered or sloping side access where the boundary exceeds the limit of construction. This is in effect a lowering of the ground to a reduced level, constructing up to former ground level and backfilling around the

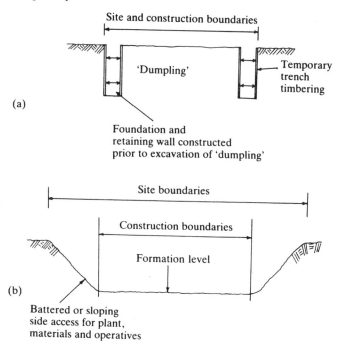

Fig. 3.17 Basement excavation (a) Perimeter trench technique
(b) Alternative without boundary restrictions

perimeter. This latter method is the simpler in theory, but may prove a problem if the water table is high. Also it is essential to check the stability of the battered sides for safe access and egress of plant and equipment.

Construction

Basement walls must be constructed with sufficient strength to retain the surrounding ground pressures in addition to transferring the superimposed structural loading to the foundations. Reinforced retaining walls shown in Fig. 3.18 containing hollow dense concrete blocks are a popular choice in excavations not exceeding 3 m depth. Walls and floor must be designed to resist penetration of moisture by provision of a waterproof membrane known as 'tanking'. In addition to 'tanking' it is necessary to ensure water tightness at construction joints in concrete, by use of stainless steel or plastic strip water bars. These are shown in place between *in situ* concrete foundation, wall and floor in Fig. 3.19.

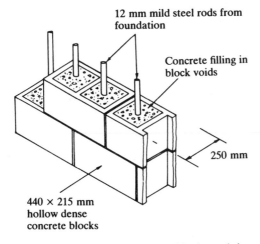

Fig. 3.18 Reinforced concrete block retaining wall

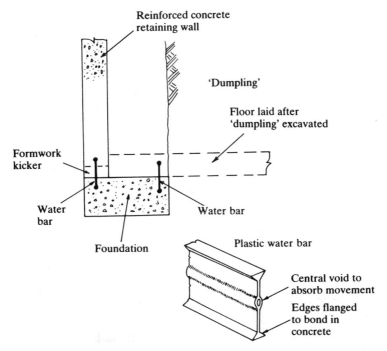

Fig. 3.19 Concrete basement wall construction

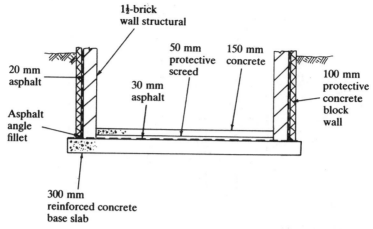

Fig. 3.20 Brick basement construction with asphalt tanking

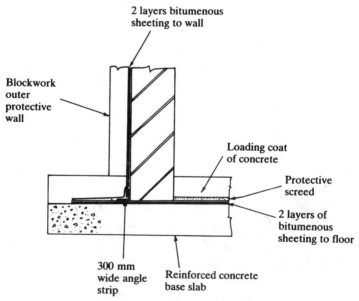

Fig. 3.21 Bituminous felt tanking to shallow basements

Tanking

Asphalt is the most acceptable waterproof membrane of 30 mm overall thickness to the floor and 20 mm to the walls. Application is in three layers for simplicity of use, and to reduce the possibility

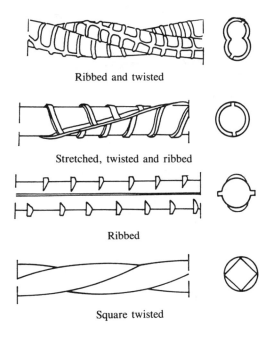

Ribbed and twisted

Stretched, twisted and ribbed

Ribbed

Square twisted

Fig. 3.22 Deformed steel reinforcement

of internal shrinkage cracking which could otherwise occur. Corners are reinforced with a 50 mm angle fillet as shown in the brick basement example in Fig. 3.20. An alternative tanking for shallow basements may be a Class A fibre-reinforced bituminous sheet membrane to BS743. The base structural surface receives a priming coat of bituminous solution before the first layer of sheet is bonded in hot bitumen, allowing 100 mm for side laps and 150 mm for end laps. Fig. 3.21 shows the use of bituminous sheet to the junction between wall and basement floor with increased thickness to the corner.

Concrete cover to reinforcement

In all steel reinforcement to concrete sub-structural work it is essential to obtain the specified minimum cover. This will be at least 40 mm preferably, 50 mm. This is achieved with spacers of concrete or plastic of the type shown in Figure 3.22.

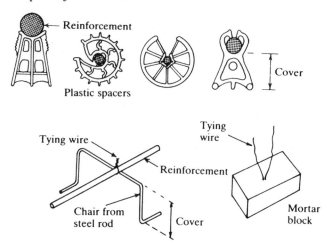

Fig. 3.23 Reinforcement spacers

Reinforcement

Steel reinforcement is twisted and deformed as shown in Fig. 3.22, to maximise its bonding potential to surrounding concrete. Additionally, it is essential to obtain adequate cover to protect the steel from dampness. This is normally 25 mm, but in sub-structural work it should be 40 mm, or preferably 50 mm. This is achieved with spacers of concrete or plastic of the type shown in Fig. 3.23.

4

Walls

Wall classification and design conveniently divide into two categories; external and internal construction. Most external walls support the upper floor and roof and most internal walls are self-supporting only, functioning as a means of dividing space. The exceptions are illustrated in Fig. 4.1, indicating cross-wall construction where the party or separating walls convey roof and

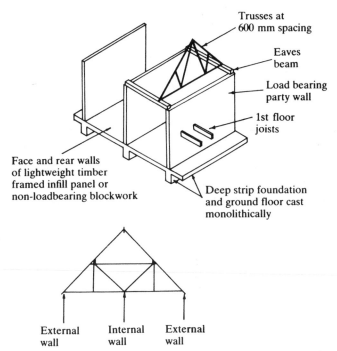

Trusses at 600 mm spacing

Eaves beam

Load bearing party wall

1st floor joists

Face and rear walls of lightweight timber framed infill panel or non-loadbearing blockwork

Deep strip foundation and ground floor cast monolithically

External wall   Internal wall   External wall

Fig. 4.1 (a) Cross wall construction (b) Traditional trussed purlin frame and roof support (see also Fig. 6.11)

floor loads, and a trussed purlin roof frame which depends on central span support from an internal dividing wall.

Strength and stability

The minimum compressive strength of individual units of construction acceptable to Building Regulation standards are 5 N/mm^2 and 2.8 N/mm^2 over gross sectional area for bricks and blocks respectively. For more detail of individual materials for brick and block manufacture the following British Standards should be consulted:

> BS 6073 – *Precast concrete masonry units*
> BS 3921 – *Clay bricks and blocks*
> BS 187 – *Calcium silicate bricks.*

The strength of mortar is a very important consideration. It should successfully join bricks while maintaining a constant separation without failure. Mortar is weaker than the associated brickwork as this is easier to replace if there is a structural defect in the wall. A composition of 1 part cement, 1 part lime and six parts fine aggregate or 1 part masonry cement to 4½ parts fine aggregate is normal for most external conditions, and 1 part cement to 3 parts sand for substructural work and very exposed situations, e.g. parapets and chimneys.

Pointing and jointing

Most brickwork joints are filled and finished in the same mortar. This is more economical than pointing, which involves raking out 15 to 20 mm of unset mortar from the joint and replacing it with a specially coloured or textured mortar. The purpose of pointing is to provide continuity of colour and economy of use with the more expensive coloured cements and colouring pigments. Pointing is also required as remedial treatment to joints in existing brick and stone walls which have suffered the effects of frost damage and settlement cracking. Examples of finishing to brickwork are shown in Fig. 4.2.

Units of construction

The units employed for construction of walls are either, brick, block or stone masonry.

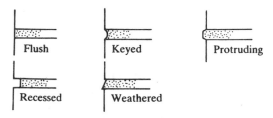

Fig. 4.2 Finishing treatment to brickwork mortar joints

Bricks

Bricks are manufactured from burnt clay, autoclaved sand and lime or coloured concrete. The standard size 215 mm × 102.5 mm × 65 mm is shown in Fig. 4.3 compared with the format size which includes a 10 mm allowance for jointing. Metric bricks are also available for dimensionally co-ordinated building, but these are less popular with bricklayers. Also, brick manufacturers and industry have been unable to find a compatible standard, therefore the following variations in format size are available:

Length (mm)	Width (mm)	Height (mm)
300	100	100
200	100	100
300	100	75
200	100	75

Actual size is 10 mm less, 12 mm for the 300 mm dimension.

Special bricks

Special applications and feature work require bricks to be reduced in size or reshaped. Specials are either cut from a whole brick as shown in Fig. 4.4 or purpose-made by hand in hardwood moulds.

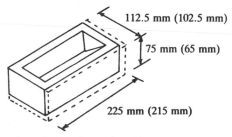

112.5 mm (102.5 mm)

75 mm (65 mm)

225 mm (215 mm)

Fig. 4.3 Standard brick format size (actual size)

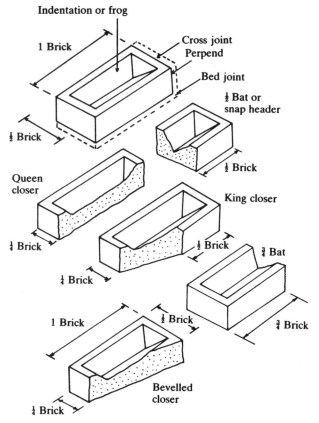

Fig. 4.4 Brick terminology and dimensions

Some examples of manufactured specials are shown in Fig. 4.5.

Bonding

The courses or rows of bricks in a wall are arranged to ensure that each brick overlaps a portion of the brick immediately below. The amount of overlap and part of the brick used determines the pattern or bond of brickwork. Bonding is provided mainly to distribute vertical and horizontal loads over a large area, to minimise movement between bricks. A secondary reason for bonding is appearance, and the examples illustrated in Fig. 4.6 show stretcher bond for normal half brick wall and outer leaf of cavity wall construction, English bond for one-brick thick structural walls, e.g. party wall, inspection chamber wall, and

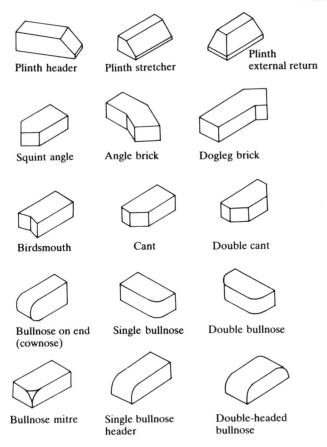

Plinth header · Plinth stretcher · Plinth external return

Squint angle · Angle brick · Dogleg brick

Birdsmouth · Cant · Double cant

Bullnose on end (cownose) · Single bullnose · Double bullnose

Bullnose mitre · Single bullnose header · Double-headed bullnose

Fig. 4.5 Purpose-made and special bricks

Flemish bond which is weaker than English bond but has a more attractive appearance. Where it is necessary to match new cavity wall construction with old solid one-brick thick (or more) walls, half bats or snap headers may be used to create the desired traditional effect, whilst still achieving modern insulation standards. Many bonding variations are possible, some are illustrated in Fig. 4.7.

Blocks

Concrete blocks are manufactured from Portland cement and aggregates. The range of aggregates used is extensive in order to achieve a variety of performance characteristics. Classification of

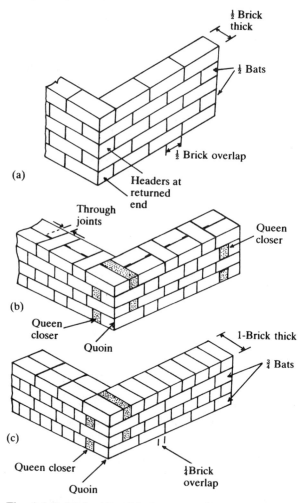

Fig. 4.6 Brick bonding (a) Stretcher bond (b) Flemish bond
(c) English bond

blocks is by compressive strength categories, 2.8, 3.5, 5, 7, 10, 15, 20 and 35 N/mm^2. The denser, stronger blocks may be used for substructural work or party wall construction in place of bricks and the lighter, weaker, hollow blocks for non-load bearing internal partition walls. The majority of blocks are 440 mm long × 215 mm high × 100 mm thick to relate to brickwork. Size range and block types are shown in Fig. 4.21.

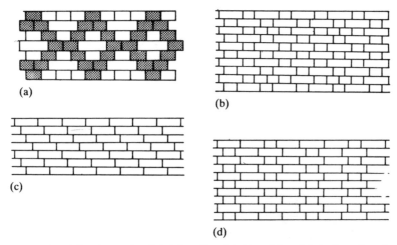

Fig. 4.7 Alternative brick bonding (a) Dutch bond (b) Cross bond
(c) Raking bond (c) Flemish garden wall bond

Stone masonry

Choice of stone is generally limited to availability in the construction area. Preponderance of natural stone deposits in some parts of the country is obvious from its abundant use as external wall cladding in these areas. Classes of building stone include:

1 Igneous rock – formed from volcanic deposits, e.g. granite, basalt.
2 Sedimentary rock – disintegrated rock, reformed by thousands of years weathering, e.g. sandstone, limestone.
3 Metamorphic rock – disintegrated rock, reformed by pressurisation or heat, e.g. marble, slate.

Reconstituted or artificial stone is also available. This is a concrete of natural stone aggregates and cement moulded into convenient size blocks. It is a substitute for natural stone and has the advantage of freedom from defects. Single-size blocks are available in addition to walling units composed of several sizes to relieve uniformity and to ensure efficient bonding.

Bonding

Stonework may be coursed by dressing the stones to an agreeable

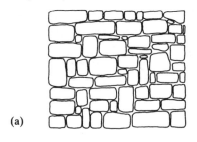

(a)

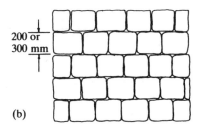

200 or
300 mm

(b)

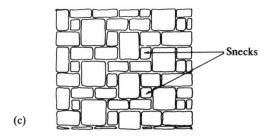

Snecks

(c)

Fig. 4.8 Variations on rubble walling (a) Random rubble
uncoursed (b) Squared random rubble, coursed
(c) Squared random rubble with through snecks

size of about 200 mm or 300 mm square as detailed in Fig. 4.8.
Alternatively, walls may be constructed from stones as they arrive
from the quarry. Awkward corners are removed and the result,
known as random rubble, uncoursed, is also shown in Fig. 4.8.
Snecked rubble walling is a compromise, composed of squared
stones of irregular size with long vertical joints interrupted by
small square stone 'snecks' of 50 mm minimum dimensions.

External walls

The stability requirements for external walls are also intended to apply to separating walls, i.e. party walls between terraced or semi-detached dwellings.

Half-brick walls (and 90 mm min. thickness blocks)

The Building Regulations make specific reference to stability requirements for non-habitable external structures such as garages, porches, sheds, etc. These apply to walls exceeding 2.5 m both horizontally and vertically providing the wall does not exceed 9 m in the direction of roof span and 3 m in height. Stability is ensured by providing 190 mm minimum square piers or 90 mm minimum buttressing at intervals not exceeding 3 m. Figure 4.9 interprets these requirements and shows measurement of wall height to gable ends.

One-brick walls (and 190 mm min. thickness blocks)

These are absolete construction for external walls in the UK, but

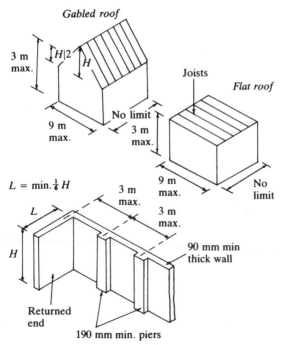

Fig. 4.9 Half-brick walls to single-storey structures

still retained for party walls between separate dwellings. Lengths exceeding 3 m must be laterally restrained by strapping, similar to the cavity detail shown in Fig. 4.14 or by end bearing of joists at floor and ceiling levels. The following table indicates the relationship between wall thickness, height and length.

Solid wall	Minimum thickness Cavity wall	Uncoursed stonework	Maximum height	Maximum length
(mm)	*(mm)*	*(mm)*	*(m)*	*(m)*
190	180	265	3.5	12
190	180	265	9	0

Note: The thickness of these walls must never be less than $\frac{1}{16}$ of the storey height which contains them.

Cavity walls

Conventional cavity walls contain a half-brick outer leaf and a 90 or 100 mm lightweight loadbearing concrete block innter leaf. Load-bearing blockwork may be used for both leaves with external surface cladding of tiles or timber to improve appearance. Maximum heights and lengths are shown in the previous table, with stability of each leaf enhanced with wall ties to BS 1243. Wall ties may be produced from galvanished steel, stainless steel or plastic, a selection is shown in Fig. 4.10. Spacing is at 900 mm maximum horizontally, 450 mm vertically and staggered as shown in Fig. 4.11, with a maximum of 300 mm (normally 225 mm to suit block dimensions) at door and window jambs as detailed in Fig. 4.11. Cavity width is normally between 50 and 75 mm, but may be up to 150 mm.

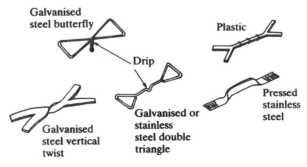

Fig. 4.10 **Wall ties**

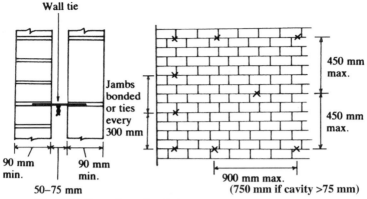

Fig. 4.11 Location of wall ties

Lateral restraint

For house construction not exceeding two storeys, walls over 3 m long must receive restraint from adjacent floor and roofs, by:

(a) Joists bearing 90 mm (75 mm with wall plate) on masonry (see Fig. 4.12); or

(b) Joists bearing on restraint type hangers, (see Fig. 4.13); or

(c) Galvanished steel straps, of 30×5 mm cross-section spaced at not more than 2 m (see Figs. 4.14, 4.15 and 6.18).

Note: (a) and (b) assumes joists spacing of not more than 1.2 m.

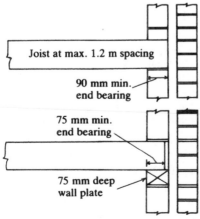

Fig. 4.12 End bearing of joists to provide lateral restraint

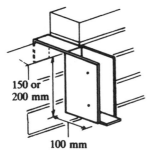

Fig. 4.13 Joist hanger acceptable as a lateral restraint tie

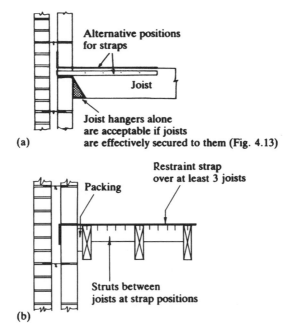

Fig. 4.14 Lateral restraint using steel straps (see Fig. 4.15)
(a) Joists square to wall (b) Joists parallel to wall

Thermal insulation

Current legislation relating to energy conservation in dwellings, requires external walls to have a maximum 'U' value of 0.45 W/m² °C. The constructional forms in Table 4.1 provide some comparison, with the reservation that thermal insulation values can deviate from those suggested with the conductive properties and quality of the components making up the structure.

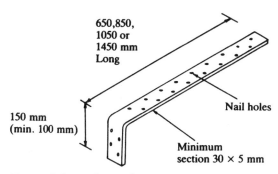

Fig. 4.15 Lateral restraint strap

Table 4.1

Wall construction	Approximate U value	
1. Solid one brick thick	2.15	
2. Cavity, brick inner and outer leaf	1.48	
3. Cavity, aerated block inner leaf and brick outer leaf	0.90	
4. Cavity, aerated block inner and outer leaf	0.71	
5. (3) with 50 mm insulation batts built into cavity	0.45	see Fig. 4.16
6. Half-brick outer leaf, 80 mm insulated timber frame inner leaf	0.40	see Fig. 4.16.

Methods of insulating conventional masonry cavity wall construction may include the following:

1. Insulation batts or slabs of polyurethane foam, expanded polystyrene or bonded glass or rock fibres. These are designed to fit tightly between wall ties at the normal 450 mm vertical spacing or 400 mm with metric bricks. Figure 4.17 shows application to new walls during construction and a method of improving an existing wall.

2. Mineral wool pellets/granules. These are pressure-blown through a flexible tube into the cavity as the work proceeds. Existing walls may receive this treatment through 65 mm diameter holes at 2.5 m maximum spacing. Voids are made good after injection of granules, with specially cut brick cores.

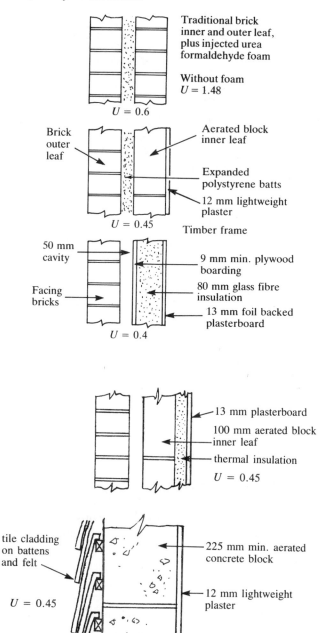

Traditional brick
inner and outer leaf,
plus injected urea
formaldehyde foam

Without foam
$U = 1.48$

$U = 0.6$

Brick
outer
leaf

Aerated block
inner leaf

Expanded
polystyrene batts

12 mm lightweight
plaster

$U = 0.45$ Timber frame

50 mm
cavity

9 mm min. plywood
boarding

Facing
bricks

80 mm glass fibre
insulation

13 mm foil backed
plasterboard

$U = 0.4$

13 mm plasterboard

100 mm aerated block
inner leaf

thermal insulation

$U = 0.45$

tile cladding
on battens
and felt

225 mm min. aerated
concrete block

$U = 0.45$

12 mm lightweight
plaster

Fig. 4.16 Wall insulation

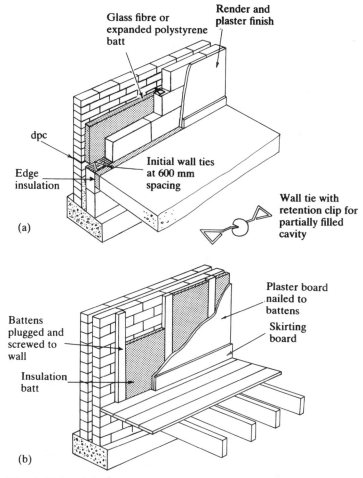

Glass fibre or
expanded polystyrene
batt

Render and
plaster finish

dpc

Edge
insulation

Initial wall ties
at 600 mm
spacing

Wall tie with
retention clip for
partially filled
cavity

(a)

Battens
plugged and
screwed to
wall

Insulation
batt

Plaster board
nailed to
battens

Skirting
board

(b)

Fig. 4.17 Insulation of brick cavity walls (a) Cavity insulation
(b) Lining

3. Urea formaldehyde foam. This too, can be applied during construction, although it is more common as a treatment to existing walls. 20 mm diameter holes are bored through the outer or inner leaf (outer preferable on existing property) at 1100 mm maximum spacing horizontally and 900 mm maximum vertically. Timber pegs are loosely placed in the holes to indicate movement as the pressure-injected foam spreads. With existing property, openings for air bricks and at the cavity head must be checked as sealed. For maximum efficiency, foam density should be about

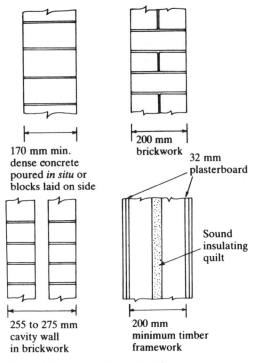

170 mm min.
dense concrete
poured *in situ* or
blocks laid on side

200 mm
brickwork

32 mm
plasterboard

Sound
insulating
quilt

255 to 275 mm
cavity wall
in brickwork

200 mm
minimum timber
framework

Fig. 4.18 Party wall construction

8 Kg/m^3. Significantly less than this will lead to generation of shrinkage cracks which will encourage water to bridge the cavity.

Sound insulation

Insulation against transmission of sound through external walls is accepted as adequate if the structural requirements are satisfied. Internal walls within a dwelling do not have to be sound resisting, but party walls dividing semi-detached and terraced dwellings must have a resistance to airborne sound which compares with the performance of a solid wall of brick or concrete having a mass of 415 kg per square metre of wall. This figure is based on the effect of the traditional one brick thick party wall. Acceptable alternatives are shown in Fig. 4.18 with the possibility of cavity construction for greater resistance to impact sound transmission.

Internal walls

Internal walls are principally to divide the gross floor area of a building into compartments or rooms. A secondary function may be to transmit floor and/or roof loads to a suitable foundation. The constructional form for partitions may be:

1   Concrete block.
2   Timber frame or stud.
3   Demountable frame.

Bricks may also be used for partitions if a high standard of fire resistance is required. They are useful if a brick feature is required, but for plastering, concrete blocks offer considerable savings.

Concrete block partitions

For load-bearing walls in single and two-storey housing the minimum thickness of blockwork is 90 mm, for three storeys, 140 mm. Stability is achieved at the base by an independant strip foundation or a thickened area of ground floor slab with reinforcement as detailed in Fig. 4.19. Where the wall only occurs in an upper storey, base support is achieved by a steel beam as shown in Fig. 4.20. This transfers the load to piers built into adjacent flank walls. Stability at the ends of block walls is shown in Fig. 4.21 with metal ties or alternative courses bonded into the inner leaf. Restraint at floor and ceiling level is provided by joists running square to the wall with at least 90 mm bearing as shown in Fig. 4.22 or by steel straps secured to at least two parallel joists, also shown in Fig. 4.22.

Non-load-bearing block partitions are less strictly controlled and may be the minimum British Standard thickness of 60 mm. Maximum height with regard to stability should not exceed 40 times the thickness inclusive of render and plaster finish. Blocks

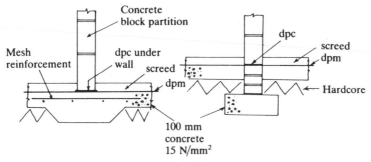

Fig. 4.19 Ground floor support to block partitions

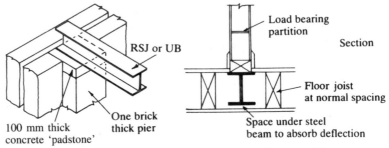

Fig. 4.20 Steel beam support to upper floor partition

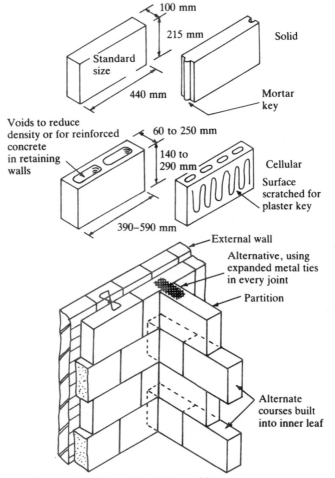

Fig. 4.21 Concrete blocks and bonded partition

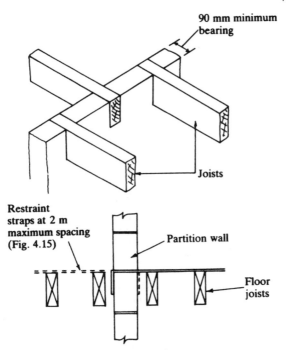

Fig. 4.22 Partition wall restraint

for this purpose will be of the minimum BS density classification and should not require a foundation in excess of the ground floor concrete. Wall ends are bonded or tied as shown in Fig. 4.21. Head restraint is advisable, and the use of timber joists and/or straps in accordance with Fig. 4.22 should be implemented. Upper floor support may be provided by a timber beam composed of two joists bolted together at half-metre intervals with toothed ring connectors sandwiched between adjacent joists. Figure 4.23 details this, and provision for door openings with continuity of jambs from floor to ceiling to improve stability. Openings in load bearing or dense concrete block partitions with require a lintel at the door head.

Timber frame or stud partition

Timber stud partitions are a lightweight wall system, generally non-load bearing and very popular for partitioning the upper floor space into bedrooms. They are constructed directly from the floor and will require no special structural support unless they are specifically designed to transmit some of the roof loading. This is

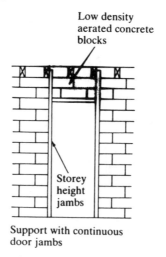

Low density
aerated concrete
blocks

Storey
height
jambs

Support with continuous
door jambs

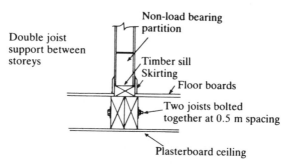

Non-load bearing
partition

Double joist
support between
storeys

Timber sill
Skirting
Floor boards

Two joists bolted
together at 0.5 m spacing

Plasterboard ceiling

Fig. 4.23 Upper floor partition support

unlikely in modern construction, but may occur in older buildings
where the roof structure requires propping at mid span. The
framed construction shown in Fig. 4.24 contains vertical studding
at 400 to 600 mm spacing with noggings at approximately 1 m
spacing to restrain movement. Noggings are staggered to simplify
nailing through the studs and door openings are provided with
thicker studs to form jambs or posts. The framework is clad with
sheet material or timber boarding. Plasterboard of 9.5 mm
thickness is the most popular, offering economy with choice of
painting, plaster or paper hanging for finish treatment. Sound or
thermal insulation may be improved by filling the framework gaps
with insulation batts as shown in the door opening detail in
Fig. 4.25.

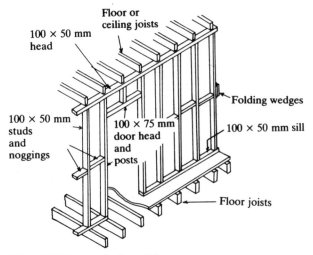

Fig. 4.24 Timber stud partition

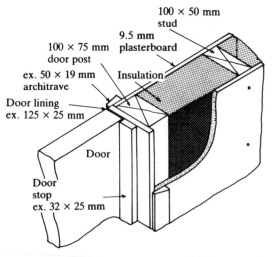

Fig. 4.25 Insulated stud partition with door opening

Demountable frame

Demountable frames are a non-load bearing partition scheme suitable for use in offices and commercial buildings. They suit this type of building, as changes in office layout or changes in occupancy can easily be achieved without structural disruption. Many patented systems exist, the type illustrated in Fig. 4.26 is

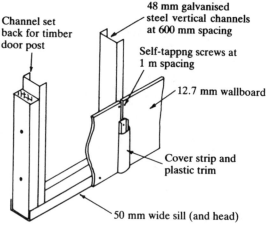

Fig. 4.26 Demountable steel-framed partition

based on a framework of lightweight galvanised steel channel fixed to wall, ceiling and floor with plugs and screws. Wallboard of plaster, chipboard or plywood is secured by self-tapping screws at approximately 1 m vertical spacings to the channels and intermediate studs spaced every 600 mm. Joints between boards are closed with a steel cover strip secured every 250 mm and a plastic capping trim.

Openings in walls

Openings in internal partition walls are required for access and privacy. The details in Figs 4.24 and 4.25 show provision in timber stud-framed partitions. In block or brick walls, support for masonry over the openings is provided by a beam or lintel of concrete or steel. Timber may also be used, but this is rare in modern constructions as quality hardwood is very expensive and softwood is suspect in the event of a fire. Lintels are subject to both compressive and tensile stresses as illustrated in Fig. 4.27. Concrete is sufficiently strong to resist the compressive force but will require the supplementary strength of steel reinforcement rods in the lower portion to resist tension. At least 25 mm protective cover is provided for the steel, to resist the effects of corrosion and fire, and end hooks are used to increase the bond between the two materials.

Prestressed concrete lintels are popular for internal walls as they

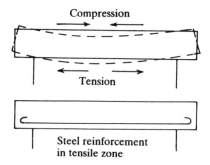

Fig. 4.27 Concrete lintel

are relatively thin and light in weight. They are manufactured by casting concrete around stretched high tensile steel wires. When the concrete has gained strength the wires are released, inducing compression throughout the concrete. This effectively provides a unit which is considerably stronger than the equivalent size of reinforced concrete, but to perform correctly they are designed to receive three or more courses of brickwork which acts with the lintel to provide a composite unit. Non-composite pre-stressed lintels are also produced, for use with very heavy loads or where there is insufficient space for superimposed brickwork. Figure 4.28 shows the construction of a prestressed lintel and the use of a corrugated galvanised lintel for internal openings. These too are a composite unit requiring at least 150 mm of brick or blockwork above the lintel.

Arches

An arch is a structure composed of several small units of brick or stone, used as an alternative to a lintel to support the load over an opening. Arch shapes may relate to many attractive geometric forms, the most common include semicircular, semielliptical and segmental. Stone is cut into wedge shapes known as voussoirs. Brick arches are either rough, axed or gauged. Rough bricks are plain and uncut, producing wedge-shaped mortar joints. For partition walls this is acceptable if the surface is plastered to disguise the construction, but where arch construction is exposed, bricks must be cut (axed) to a taper to produce parallel mortar joints. Where the arch effect is a very special feature, bricks are cut and ground (gauged) to provide an extremely precise and fine joint, usually in white mortar. Support for arches during

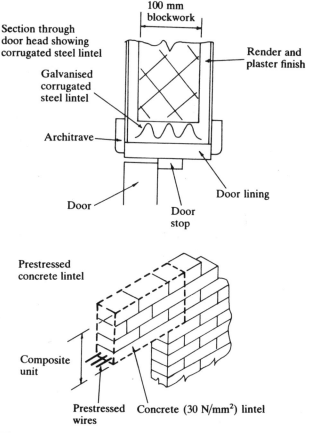

Section through
door head showing
corrugated steel lintel

100 mm
blockwork

Render and
plaster finish

Galvanised
corrugated
steel lintel

Architrave

Door

Door
stop

Door lining

Prestressed
concrete lintel

Composite
unit

Prestressed
wires

Concrete (30 N/mm^2) lintel

Fig. 4.28 Internal wall lintels

construction is from a centre, or a turning piece for low rise arches. A centre is produced from timber, cut to follow the arch profile, with laggings shown in Fig. 4.29 providing direct support for the bricks. The turning piece, also shown in Fig. 4.29, is suitable for a segmental arch in a partition wall; it includes the use of struts and folding wedges for easy removal of the support without damaging the arch.

Openings in cavity walls

Openings are required as provision for doors and windows. These effectively bridge the cavity, therefore means of preventing the

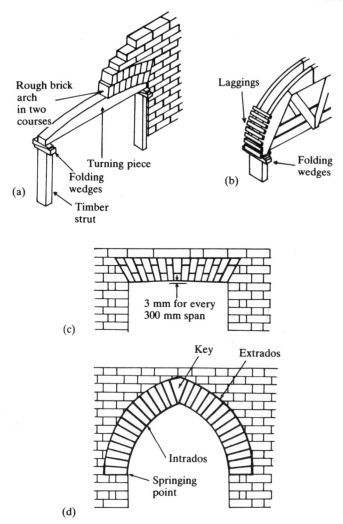

Fig. 4.29 Arch construction (a) Segmental arch (b) Arch centre
(c) Gauged arch (d) Mitred arch

passage of rainwater access across constructional openings must be
incorporated in the unit design. Door and window sills have a
sloping or weathered edge, with drip or groove cut in the
underside. Weather bars are another important feature between
door and sill, and anti-capillarity grooves a necessity between sash
and window frame. Figures 4.30 and 4.31 illustrate typical

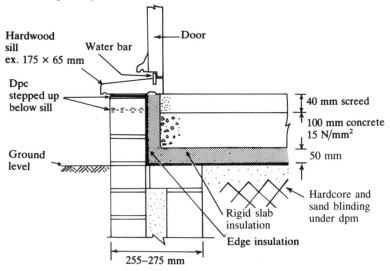

Hardwood sill ex. 175 × 65 mm

Water bar

Door

Dpc stepped up below sill

40 mm screed

100 mm concrete 15 N/mm²

50 mm

Ground level

Rigid slab insulation

Edge insulation

Hardcore and sand blinding under dpm

255–275 mm

Fig. 4.30 Threshold construction

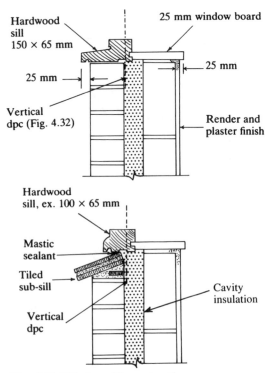

Hardwood sill 150 × 65 mm

25 mm window board

25 mm

25 mm

Vertical dpc (Fig. 4.32)

Render and plaster finish

Hardwood sill, ex. 100 × 65 mm

Mastic sealant

Tiled sub-sill

Vertical dpc

Cavity insulation

Fig. 4.31 Window sill construction

construction at threshold and window sill, incorporating the weather exclusion features just detailed.

The construction at the side or jambs of doors and windows requires a solid wall, as shown in Fig. 4.32. The inner leaf of blockwork is returned to meet the brick outer leaf and a vertical damp-proof course is used between block and brickwork to

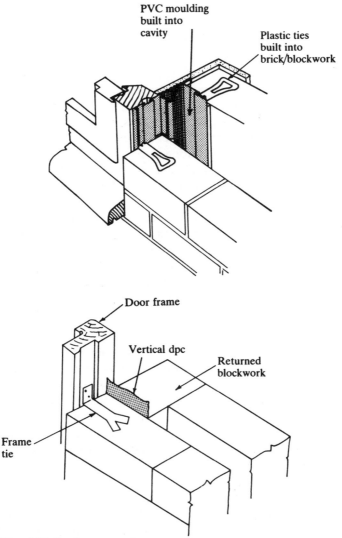

Fig. 4.32 Jamb construction

prevent dampness bridging. The jamb is attached to the brickwork with fish-tailed galvanised steel ties at 300 mm intervals for doors and 450 mm for windows. Where the cavity is insulated, the returning blockwork could cause a cold bridging effect. In these circumstances it may be preferable to continue the insulation up to the back of the frame using an extruded plastic cavity closer, also shown in Fig. 4.32.

Lintels

A lintel is a beam providing support to the superstructure above a wall opening. Reinforced concrete has enjoyed considerable use but, over the last twenty years, has gradually been replaced with lighter weight galvanised steel sections. Composite units of steel reinforcement and concrete are very heavy and difficult to handle, and for larger spans will have to be cast *in situ*. Also, they require a dpc and are unsightly, leaving an exposed section on the face of walls, unless provided with a steel angle to support the outer leaf as shown, with another variation in Fig. 4.33. Both concrete and steel lintels incorporate a sloping face or stepped damp-proof course to deflect any dampness occurring in the cavity outwards.

Galvanised steel lintels are either a single pressed section or boxed, the latter filled with insulation if required. They have excellent compressive and tensile strength properties and a weight which is only a fraction of a comparable concrete unit. Most units have the dpc incorporated in the galvanised surface finish, others as compared in Fig. 4.34 require a separate dpc stepped above the lintel.

Scaffolding

Scaffold and ladders are required for access mainly to construct walls, upper floors and roof when brickwork exceeds about chest level above the ground. Since the enforcement of the Health and Safety at Work, etc. Act, the requirements and regulations associated with safe working on scaffolding systems have intensified considerably. For details the reader should refer to BS 5973 Code of practice for access and working scaffolds. This section is intended to provide no more than an introduction to scaffolding, with basic requirements for correct use.

Systems

Scaffold systems contain timber boards to BS 2482 and metal tube to BS1139. Boards should be free of defects, such as knots and

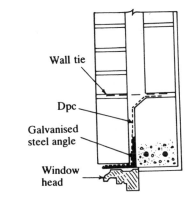

(a)

Wall tie

Dpc

Galvanised
steel angle

Window
head

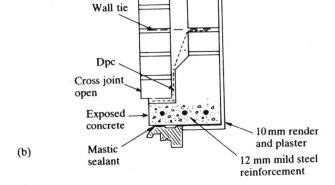

(b)

Wall tie

Dpc

Cross joint
open

Exposed
concrete

Mastic
sealant

10 mm render
and plaster

12 mm mild steel
reinforcement

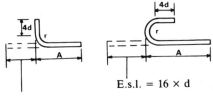

(c)

E.s.l. = 16 × d

Equivalent straight length
= 8 × bar diameter, d.

Fig. 4.33 Door/window head construction with reinforced concrete
lintels (a) Soldier arch (b) Boot lintel (c) End treatment
to reinforcement

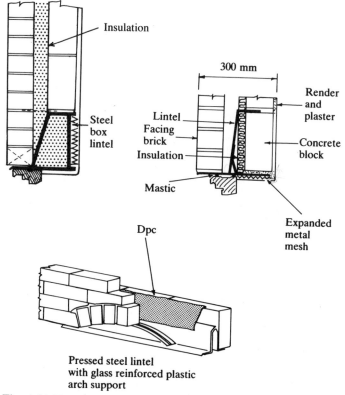

Fig. 4.34 Door/window head construction with pressed steel lintels

deformities and have ends protected with galvanised steel hoops. Standard width is 225 mm and thickness 38 mm. Metal tube may be black or galvanised mild steel or aluminium alloy, all with an outside diameter of 48.4 mm.

Putlog scaffold (Fig. 4.35)

This system builds into the brickwork as the building increases in height. Ledgers correspond with heights of lift at about 1.4 m intervals and vertical stability is achieved with only one row of standards. Putlogs bear directly on to the wall. Only the highest lift may be used as a working platform and, in the interests of safety, the maximum height should not exceed three storeys. Where putlogs support boards, at least three are required at a maximum spacing of 1.5 m (38 mm boards). Standards should be not more than 1.8 m apart, positioned 1.3 m from the wall to allow a

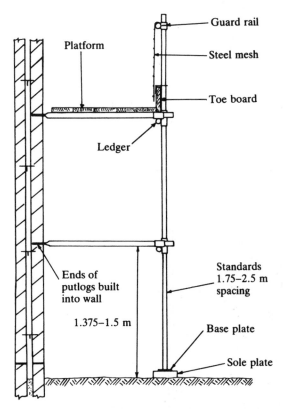

Guard rail

Platform

Steel mesh

Toe board

Ledger

Standards
1.75–2.5 m
spacing

Ends of
putlogs built
into wall

1.375–1.5 m

Base plate

Sole plate

Fig. 4.35 Putlog scaffold

five-board working platform. At each lift one putlog must remain within 300 mm of every standard.

Independent scaffold (Fig. 4.36)

This system has two rows of standards at 2.1 m to 2.7 m spacing depending on loading and height. It is ideal as access for maintenance to existing buildings and as access during construction to high rise buildings. Heights of lift or ledger levels will occur at about 1.4 m for bricklaying and up to 3 m for multi-storey construction.

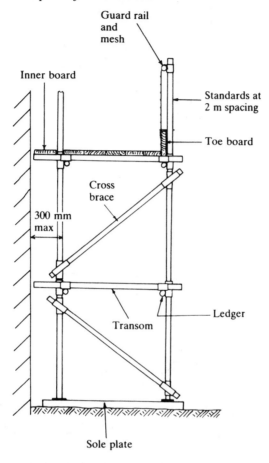

Fig. 4.36 Independent scaffold

Platforms (putlog and independent)

The requirements for safe working on scaffold platforms are shown in Fig. 4.37. This detail includes legislation for toeboards, handrail, mesh and ladders.

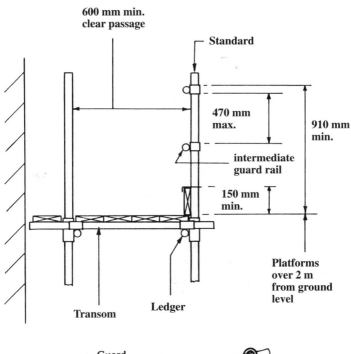

600 mm min. clear passage

Standard

470 mm max.

910 mm min.

intermediate guard rail

150 mm min.

Platforms over 2 m from ground level

Transom

Ledger

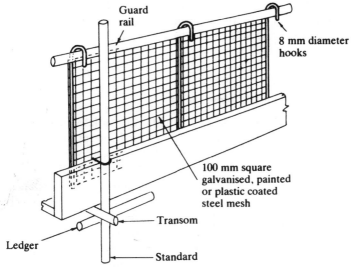

Guard rail

8 mm diameter hooks

100 mm square galvanised, painted or plastic coated steel mesh

Transom

Ledger

Standard

Fig. 4.37 Scaffold platform

5

Floor construction

A floor structure must fulfil several functions, including the following design considerations:

1. Provision of a uniform level surface.
2. Sufficient strength and stability.
3. Exclusion of dampness from the inside of a building (ground floor).
4. Thermal insulation (max. 0.45 $W/m^2 K$).
5. Sound insulation (in flats).
6. Resistance to fire.
7. Compatibility with the surface finish.

This chapter considers these factors where appropriate to floor construction from, structural timber, *in situ* plain and reinforced concrete and precast self-centering concrete floors.

Timber

Suspended timber ground floors

Timber ground floors are now virtually obsolete, due to the escalating cost of the materials and skilled labour required for their installation. They are used where extensions correspond with the existing structure, and when specified for individually designed premises where cost limits are not restrictive. The suspended timber structure offers the benefit of comfort but, if incorrectly constructed, the problems of decay. Air circulation beneath the floor surface is essential to prevent condensation. This is achieved with air bricks in the external wall at a maximum of 2 m spacing and an air void of at least 125 mm beneath joists. Figure 5.1 shows a typical section through the floor construction with intermediate and end support to joists from perforated or honeycomb dwarf

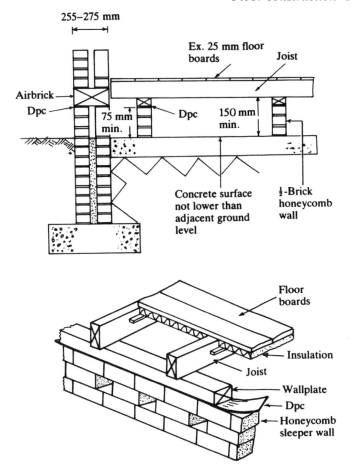

255–275 mm

Ex. 25 mm floor boards

Joist

Airbrick
Dpc

75 mm min.

Dpc

150 mm min.

Concrete surface not lower than adjacent ground level

½-Brick honeycomb wall

Floor boards

Insulation

Joist

Wallplate

Dpc

Honeycomb sleeper wall

Fig. 5.1 Suspended timber ground floor construction

walls. Damp-proof courses are provided in each support wall and insulation may suspend from joists to prevent draughts through butt-jointed boards and to reduce heat loss.

Suspended timber upper floors

Suspended timber ground floors have the convenient sublayer of concrete to provide intermediate support for joists. Spans are minimal, and joists small in comparison to upper floors where joists extend between external walls, except where partitions intercede. These are known as single floors, and an example of this type of structure is shown in Fig. 5.2. In the absence of a

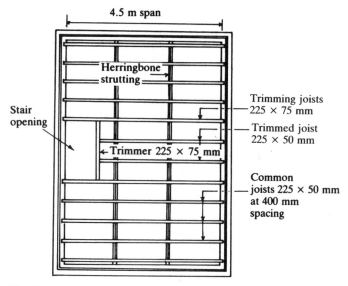

Fig. 5.2 Suspended timber upper floor (shown on plan with boards omitted)

load-bearing partition where clear ground-floor space is required, joists may be supported by a steel beam or two joists bolted together. This is double-floor construction shown with beam-to-joist connections in Fig. 5.3.

Strutting and trimming

Strutting is provided to floor joists at mid-span or 1.5 m spacing to resist buckling and deformity. The herringbone arrangement using 50 or 38 mm square softwood or thin galvanised steel struts is most efficient, but solid strutting is often used for easier and quicker installation. Between the first and last joists and adjacent walls folding wedges are used to firmly locate the strutting.

Openings in the floor structure for stairs and service ducts are provided by terminating the affected joists with a trimmer which bear on two trimming joists. The trimmer is usually 3 mm wider than the common joists, for every joist trimmed into it, and the trimming 25 mm wider than the common joists. If there is any question of these units being insufficient, their size may be checked by the design procedure explained on page 111. Figure 5.4 shows part of an upper floor construction, with boards removed to expose trimming around a stair opening with built-in end support

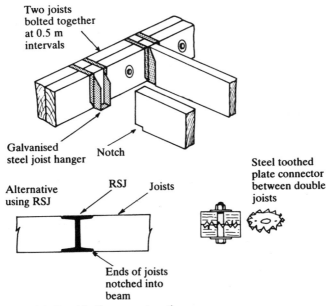

Two joists bolted together at 0.5 m intervals

Galvanised steel joist hanger

Notch

Alternative using RSJ

RSJ

Joists

Steel toothed plate connector between double joists

Ends of joists notched into beam

Fig. 5.3 Double floor construction

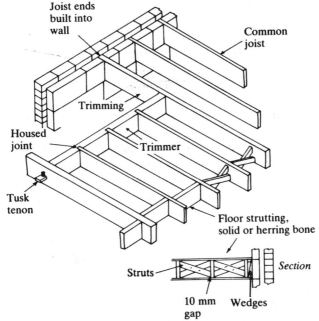

Joist ends built into wall

Common joist

Trimming

Housed joint

Trimmer

Tusk tenon

Floor strutting, solid or herring bone

Struts

10 mm gap

Wedges

Section

Fig. 5.4 Upper floor construction in timber

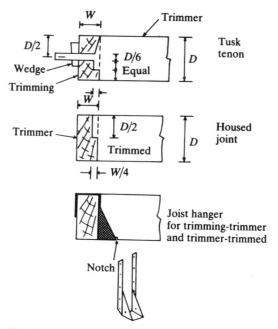

Fig. 5.5 Treatment of timber at stair openings

to joists and herringbone strutting. Figure 5.5 details the construction, showing both traditional craftsman-made joists and the use of contemporary joist hangers.

Access for services

Provision for service pipes and cables through joists is known as notching and holing. Notching is the removal of a top section of a joists as shown in Fig. 5.6. This should be of minimal width to accommodate a pipe and in particular minimal depth, otherwise the original design depth is considerably less effective. Wherever possible pipes should run parallel to joists, but this is not always convenient. Cables are more flexible and can locate through joists by holing in the neutral axis. This area occurs where compressive and tensile load distribution neutralises in the centre of a joist. Figure 5.6 also shows the areas of maximum shear force close to adjacent wall support, and the mid span area of maximum bending. Both should be avoided for service access.

Timber joist and beam design

The size of timber joists is determined by four factors:

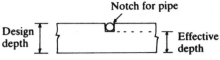

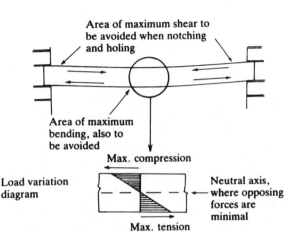

Fig. 5.6 Holing and notching joists

1. The clear span between supports.
2. The total load they carry.
3. Their spacing.
4. The quality of timber employed.

Methods for obtaining joist dimensions fall into three categories:

1. Empirical formulae.
2. Design tables.
3. Design formulae

Empirical formulae

Empirical calculations are unreliable, as they fail to incorporate variable conditions. They are not recommended as the sole method for designing joists, although they are useful for checking or comparing with the other methods. The following well-established rule of thumb, which assumes a section width of 50 mm and joist spacing of 400 mm, may be used as a guide to joist depth (D):

$$D(mm) = \frac{span\ (mm)}{24} + 50$$

e.g. if span between walls if 3.6 m

$$D = \frac{3600}{24} + 50$$
$$= 150 + 50$$
$$= 200 \text{ mm}$$

Joists are 50 × 200 mm at 400 mm spacing.

Design tables

Design tables are included with the Building Regulations. Here, the variables of span, floor load, spacing and timber quality combine. Floor dead loading is divided into three sections, ranging between 0.25 and 1.25 kN/m² and joist spacing provided with three alternatives, 400, 450 and 600 mm. The lower figure of 400 mm is preferred for compatibility with plaster ceiling boards and plywood or chipboard decking. 600 mm spacing will require uneconomically thick ceiling and floor boarding to avoid sagging between joists. Choice of timber is limited for economical reasons to softwoods. These vary considerably in quality, depending on the source of timber, the species and method of growing. In order to provide a common, easily understood standard, all softwoods for structural use are graded as Special Structural (SS) or General Structural (GS). The difference is determined by proportion of defects such as knots, wane, and deformities as defined in BS4978; Timber Grades for Structural Use. Table 5.1 is a design guide suitable for floor joists of GS grade timber. Notice that the previous empirical example is confirmed adequate, with a maximum span of 3.74 m.

Table 5.1 Floor joist design table

	Spacing (mm)		
Joist size (mm)	*400*	*450*	*600*
	Span (m)		
38 × 150	2.5	2.35	1.93
38 × 200	3.27	3.09	2.68
47 × 150	2.78	2.62	2.24
47 × 200	3.63	3.43	2.98
50 × 150	2.86	2.70	2.33
50 × 200	3.74	3.53	3.07
63 × 150	3.10	2.98	2.63
75 × 200	4.35	4.19	3.74

Floor dead load, 0.5−1.25 kN/m², imposed load not exceeding 1.5 kN/m²

Design formulae

Calculation of joist size by design formulae is necessary for unusual loading and where span or joist spacing are excluded from design tables. The procedure is as follows:

1. Summate the total load in kilograms.
2. Convert this to a unit of force in newtons by multiplying by 9.81 (use 10 for convenience).
3. Apply the bending moment formula.
4. Apply the section design formula.

The following example illustrates this process:

(1) A timber section is required to transmit a uniformly distributed total floor load of 500 kg, between opposing walls 5 m (5000 mm) apart.
(2) 500 kg × 9.81 (10) = 5000 Newtons
    Bending moment (BM) formula will depend on whether the load is at a point on the beam or whether it is distributed. Figure 5.7 shows the difference.

(3) In this example $BM = \dfrac{Wl}{8}$

   therefore $BM = \dfrac{5000 \times 5000}{8}$

   $= 3\ 125\ 000$ Nmm

(4) Design formula: $BM = \dfrac{fbd^2}{6}$

   f = fibre stress of timber, see BS 5268: Pt 2.
   b = breadth of section.
   d = depth of section.

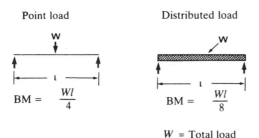

Point load   Distributed load

W

$BM = \dfrac{Wl}{4}$ $BM = \dfrac{Wl}{8}$

W = Total load
l = Effective span

Fig. 5.7 Bending moment formulae

A fibre stress of 5 N/mm² is chosen, as sufficient for most GS grade softwoods, and the section breadth is assumed to be 75 mm.

thus; $3\ 125\ 000 = \dfrac{5 \times 75 \times d^2}{6}$

transposing; $d^2 = \dfrac{6 \times 3\ 125\ 000}{5 \times 75}$

$\qquad d = \sqrt{50\ 000} = 223.6$ mm

Nearest commercial size is 75 × 225 mm.

If the depth is unsatisfactory, re-apply the formula with an alternative assumed breadth, i.e. 50 mm or 100 mm.

Timber floating floor construction

When an existing house or building is converted into flats, the intermediate floors must be constructed to a standard which prevents noise nuisance between the different occupancies. To satisfy the Building Regulations, existing timber floors must be upgraded to resist the transmission of both airborne and impact sound. This is achieved by increasing the density and resilience of the floor. Possible construction is shown in Fig. 5.8 with density improvements provided with at least 19 mm of plasterboard and 50 mm of dry sand pugging between joists. Resilience is achieved using a sound-absorbing glass fibre or mineral wood quilt draped over the existing joists, and the floor boards are supported by 50 × 50 mm battens to float over the joists. The floating surface of boards are nailed to the battens, but these nails must not connect with the joists.

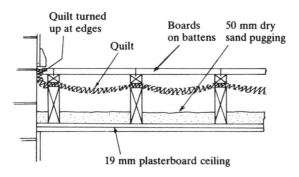

Fig. 5.8 Floating timber floor construction

Concrete

In situ ground floors

Solid concrete ground floors have four principal components; hardcore, a damp-proof membrane insulation (to achieve a 'U' value of 0·45) and a layer of dense concrete.

Hardcore is compacted onto the reduced ground level after the first 150 to 250 mm of topsoil are removed. Topsoil is soft and compressible, and contains growing matter, therefore it must be discarded. Hardcore may be brick rejects, demolition rubble or any other broken masonry of 50 to 75 mm dimensions, laid to a total thickness of at least 100 mm. Its purpose is to consolidate the ground to reduce the amount of concrete which would otherwise be required. It must be carefully selected to ensure freedom from impurities containing sulphates, which are renowned for deteriorating concrete by decomposition of the cement. To prevent cement grout loss from the superimposed concrete layer, or to protect a damp-proof membrane from fracture, the hardcore is blinded with a 25 mm layer of sand.

The damp-proof membrane may be positioned below the concrete slab, upon the sand blinding. Polythene sheet of at least 1000 gauge (0·26 mm) is the most popular material, although a LDPE radon/methane membrane of at least 0.3 mm will be essential in many areas. Alternatively, the dpm may be sandwiched between a 30 mm finishing screed and the structural concrete slab. Sheet membranes are not recommended here, unless the screed is at least 50 mm. They create a separating layer which would cause the screed to act independently, and possibly crack. In preference, cold or hot application of bituminous solution in three layers with the final layer sprinkled with sand to bond to the screed overlay. Both dpm positions and associated construction are shown in Fig. 5.9. The concrete slab is at least 100 mm in thickness, composed of cement, fine aggregates and coarse aggregates in the ratio of 1 : 3 : 6 to provide a minimum strength specification of 15 N/mm^2 at 28 days. A tamping bar is used to compact and level the concrete with finishing provided by a power float when the concrete is semi-hard or by cement and sand (1 : 3–4) screed. Figure 5.10 shows tamping bar and power float application.

In situ concrete upper floors

Unlike ground floors, concrete upper floors and flat roofs have no continuity of support. They must span from wall to wall and

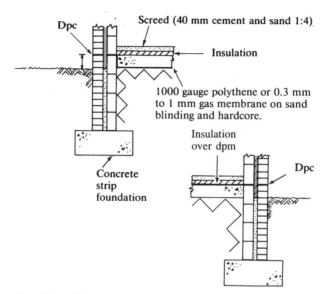

Fig. 5.9 Solid concrete ground floor construction

remain self supporting. Simple suspended concrete floors are one-way spanning only, they are designed like a continuous lintel. Two-way spanning floors incorporate the same design principles in both elevations, but these will be avoided at this level as they are most suited to large span structures.

Single span floors are at least 100 mm thick with the empirical formula of 40 mm thickness for every 1 m span, providing a guide to thickness : span ratio. Steel reinforcement is essential to resist tensile stresses in the lower portion of the slab and shear stresses close to wall support. Figure 5.11 shows alternative reinforcement arrangements which satisfy these objectives.

Reinforcement schedules

Reinforcement is conveniently co-ordinated into schedules for simplicity of design, ordering, fabrication and assembly. BS 4466 provides a coding system which clearly defines the most common reinforcement shapes, part of which is illustrated in Fig. 5.12. By using these shape codes and the abbreviated nototation shown on the reinforcement concrete upper floor slab in Fig. 5.13, steel bars can be conveniently listed in a schedule of the type shown in Fig. 5.14. The notation 3R10-2-300T translates:

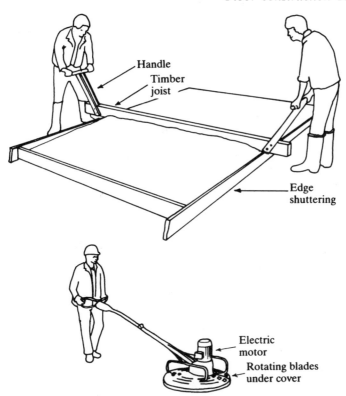

Handle
Timber joist

Edge shuttering

Electric motor
Rotating blades under cover

Fig. 5.10 Concrete finishing (a) Tamping bar (b) Power float

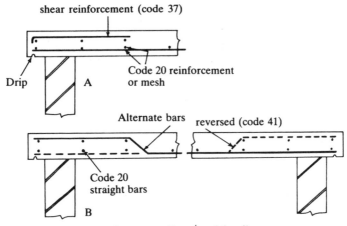

shear reinforcement (code 37)

Code 20 reinforcement or mesh

Drip   A

Alternate bars   reversed (code 41)

Code 20 straight bars

B

Fig. 5.11 Reinforced concrete floor/roof details

3 = No of bars
R = Mild steel bars
10 = 10 mm diameter
2 = bar reference mark
300 = spacing in millimetres
T = placed in Top of slab

The other variables shown, are;

T, before the diameter = high yield steel
B = placed in Bottom of slab
abr = alternate bars reversed.

Formwork to upper floors

Formwork is the temporary moulding and associated support necessary to shape concrete to the desired profile. Repetitive elements such as lintels and columns may be formed with plastic or

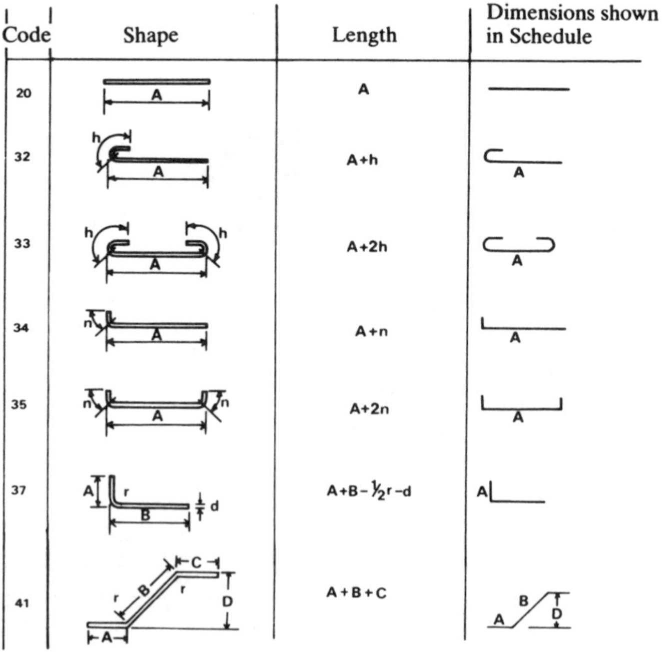

Code	Shape	Length	Dimensions shown in Schedule
20		A	
32		A+h	
33		A+2h	
34		A+n	
35		A+2n	
37		$A+B-\frac{1}{2}r-d$	
41		A+B+C	

Note: End hooks are bent to a radius,
$r = 2 \times$ bar diameter for mild steel,
and $3 \times$ bar diameter for high yield steel.

Fig. 5.12 Bar bending extract from BS 4466

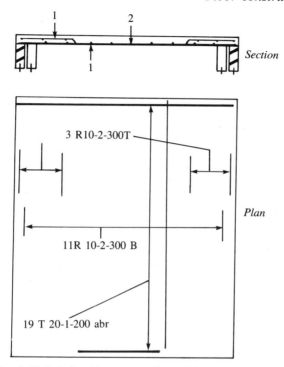

Fig. 5.13 Reinforced concrete floor slab, type B (Fig. 5.11)

steel moulds which have numerous reuses. Floors, unless repeating in several storeys, are conveniently shaped from plywood bearing on timber joists. Figure 5.15 shows a section typical of traditional formwork with double joist or ledger support to joists if load or span require it. Ground support is provided by timber or adjustable steel props. The concrete is shown here with a slight projection including associated formwork, bracketing and moulding to create a drip.

Plywood for formwork is generally specified as shuttering grade, which has one side only prepared. Where a particularly smooth finish is specified, hardboard lining may be used or plastic faced plywood. These have the added advantage of preventing the water content of concrete penetrating the plywood which will cause spalling and surface irregularities when the formwork is removed. Alternatively, exposed timber formwork should be sprayed or painted with a proprietary mould oil to provide an impervious surface film.

Member	Bar mark	Type and size	No of mbrs	No of bars in each	Total no	Length of each bar mm	Shape code	A mm	B mm	C mm	D mm	E/r mm
Pad foundation	01	T16	4	20	80	1350	35	1100				
	02	R16	4	4	16	1200	37	100				
	03	R6	4	5	20	1300	60	300	300			
Slab	1	T20	1	19	19	4740	41	4000	140	1600	100	
	2	R10	1	17	17	5000	20	5000				

Bar schedule ref:
Site ref:
Date prepared
Prepared by

20 T 1601–200 B 10 EW

4R 16 02

5R 603–200

25 N/mm²

Fig. 5.14 Bar schedule for pad foundation shown and floor slab (Fig. 5.13)

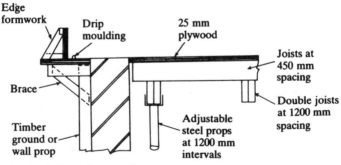

Edge formwork
Drip moulding
25 mm plywood
Brace
Timber ground or wall prop
Adjustable steel props at 1200 mm intervals
Joists at 450 mm spacing
Double joists at 1200 mm spacing

Fig. 5.15 Floor formwork

Precast self-centering floor systems

Precast floor systems are also known as self-centering or self-forming floors, as they are installed without formwork. Over large spans, there is an exception with composite (superimposed

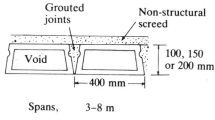

Fig. 5.16 Precast concrete floors (hollow beams)

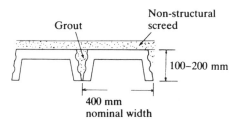

Fig. 5.17 Precast concrete floors (inverted channels)

structural topping) floors. These will require temporary propping at mid span. The other advantages are the immediate availability of a working platform without the delay waiting for concrete to gain strength, and the guaranteed quality of each factory produced unit with no concern about the effects of weather or labour problems with regard to production schedules. The disadvantages are minimal and include the need for a crane where work exceeds ground level, less design flexibility and the need for very close supervision of accuracy in constructing the frame or wall support structure in order to ensure adequate bearing and unit co-ordination.

There are a wide range of precast designs and systems, which divide into five categories:

1. Rectangular hollow cross-sections, close spaced (Fig. 5.16).
2. Inverted channel sections, close spaced. (Fig. 5.17).
3. Beams spaced apart with lightweight infill blocks between (Fig. 5.18).
4. I beams with projecting reinforcement for location to a structural topping. Beams are spaced apart for location of precast concrete panels which become permanent formwork (Fig. 5.19).

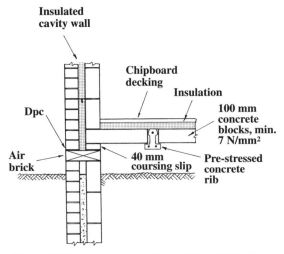

Insulated cavity wall

Chipboard decking

Insulation

100 mm concrete blocks, min. 7 N/mm²

Dpc

Air brick

40 mm coursing slip

Pre-stressed concrete rib

Fig. 5.18 Precast concrete floors (rib and block)

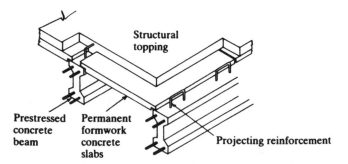

Structural topping

Prestressed concrete beam

Permanent formwork concrete slabs

Projecting reinforcement

Fig. 5.19 Precast concrete floors (prestressed beams and precast slabs)

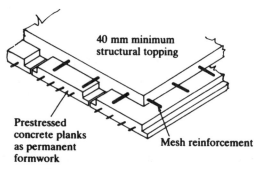

40 mm minimum structural topping

Prestressed concrete planks as permanent formwork

Mesh reinforcement

Fig. 5.20 Precast concrete floors (prestressed planks)

5. Close-spaced structural planks which also act as permanent formwork to an *in situ* structural topping (Fig. 5.20).

Note: Examples 1, 2 and 3 are known as fully precast, as an *in situ* screed or concrete topping is optional. Examples 4 and 5 are composite systems, requiring a structural concrete topping to neutralise the prestressing effects and to provide adequate depth of structure.

6

Roof construction and covering

The roof structure is designed principally to prevent peneration of inclement weather and to provide an adequate barrier against heat loss. Other considerations include an acceptable appearance, the facility to absorb thermal and moisture movement, a durable finish and sufficient strength to accommodate maintenance and snow loads.

Figure 6.1 shows a combination of roof formations. This unlikely arrangement indicates constructional forms, components and allied terminology which must be understood before proceeding with this chapter.

Roof classification

Roof structures are classed according to the interrelationship of components which make up their framework:

1. Single roofs.
2. Double roofs.

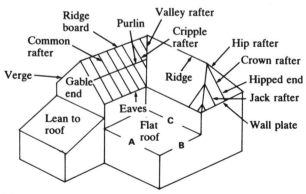

Fig. 6.1 Roof components

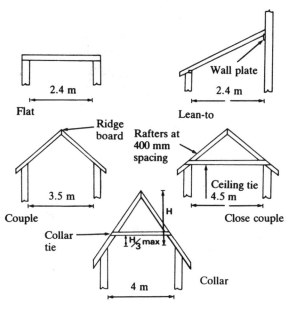

Fig. 6.2 Types of single roof

3. Triple or framed roofs.
4. Trussed rafters.

Single roofs

Single roofs are produced in a variety of forms, all having the common property of two-dimensional support, except at ridgeboard and wallplate levels. These roofs are simple in design and include the flat, lean-to, couple, close-couple and collar patterns shown in Fig. 6.2. Lack of intermediate or third dimensional support restricts loading and span potential to the figures suggested. These are based on 100 × 50 mm rafters and joists at 400 mm spacing.

Flat roofs

The term flat in this situation is inappropriate, as it implies a perfectly level surface. This is impractical for surface water disposal, so flat roofs are effectively inclined to a slope of up to 10°. The slope or fall is generally achieved by nailing firrings (long tapering wedges) to the top side of joists. The alternatives are to cut the joists to a taper, which is wasteful, or to incline the joists, which leaves a sloping soffit. Joist sizing depends on spacing, span and loading, and this may be obtained from Building Regulation tables or by design, as explained in Chapter 5. Herring-bone or solid strutting is required if the span exceeds 2.4 m and decking is provided by prefelted plywood or chipboard.

Details of possible construction at eaves (A), verge (B) and abutment (C) are shown in Figs 6.3–5. The letters A, B & C refer to Fig. 6.1. Waterproofing is shown in two layers of bituminous roofing felt to BS 747 with stone chippings

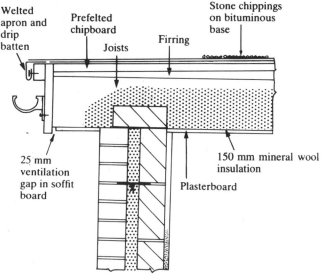

Fig. 6.3 Flat roof construction. Eaves (detail A, Fig. 6.1)

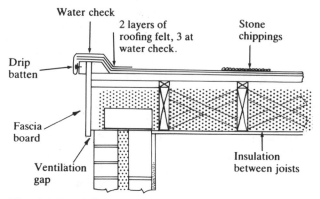

Fig. 6.4 Flat roof construction (detail B, Fig. 6.1)

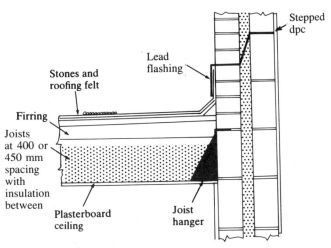

Fig. 6.5 Flat roof construction. Abutment (detail C, Fig. 6.1)

superimposed to reflect heat. Three layers of felt should be used where the decking is not prefelted or bitumen sealed. The felt is bonded with hot or cold bituminous solution and laid with staggered joints of 50 mm minimum side lap and 75 mm end lap.

Lean-to

A lean-to roof structure is a simple means of covering a small extension in preference to a flat roof, where a traditional pitched feature is required. The addition of horizontal joists will provide for a flat ceiling or loft storage space. Rafters may be built into the outer leaf of the cavity wall, but for easier levelling and fixing, a wall plate plugged and screwed to the wall will be preferred. Figure 6.6 shows possible construction to an existing wall, showing plain tiling, lead flashing and the essential use of a polypropylene cavity tray. If this tray is omitted, dampness penetrating the outer leaf could penetrate the structure below.

Couple and close-couple roofs

Couple roofs contain pairs of opposing rafters with central rigidity from a ridge board. They are relatively weak and have a consequent span restriction of about 3.5 m, limiting their application to small garages, sheds and similar single-storey buildings. The main weaknesses and limitations are deflection of

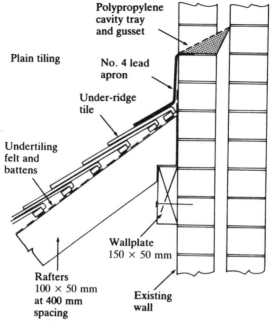

Polypropylene
cavity tray
and gusset

Plain tiling

No. 4 lead
apron

Under-ridge
tile

Undertiling
felt and
battens

Wallplate
150 × 50 mm

Rafters
100 × 50 mm
at 400 mm
spacing

Existing
wall

Fig. 6.6 Lean to roof – abutment detail

the rafters and splaying out of the support walls. This latter
problem may be resolved by using ceiling ties, which also provide
joist support for a ceiling finish. In close-couple form the span
potential is greater, but intermediate support to the rafters will be
necessary over 4.5 m. Figure 6.7 details ridge construction with
plain tiling, and Figs. 6.8 and 6.9 possible eaves treatment with

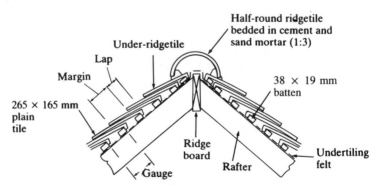

Half-round ridgetile
bedded in cement and
sand mortar (1:3)

Under-ridgetile

Lap

Margin

265 × 165 mm
plain
tile

38 × 19 mm
batten

Ridge
board

Rafter

Undertiling
felt

Gauge

Fig. 6.7 Ridge construction with plain tiles

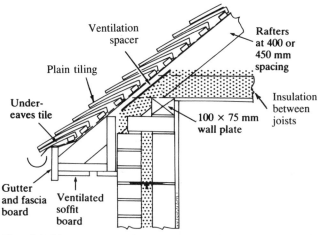

Fig. 6.8 Closed or boxed eaves construction

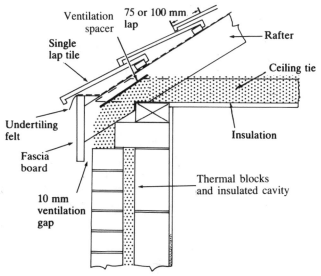

Fig. 6.9 Flush eaves construction with single lap tiling

boxed and flush finishes respectively. The flush eaves detail is shown with single-lap interlocking tiles.

Collar roof

The collar roof is a variation of a close-couple roof with the ceiling tie raised. This roof form economises in brickwork by utilising part

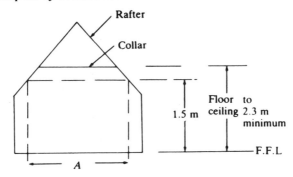

Gabled roof
A × room length = at least ½ total floor area

Hipped roof
A × similar dimension in opposing direction obtained with another 1.5
line = at least ½ total floor area

Fig. 6.10 Collar roof, construction limitations

of the roof space for accommodation. Popular applications include
chalet bungalows and loft conversions, but for living convenience
the skieling or sloping ceiling should be restricted as shown in
Fig. 6.10.

Double roofs

Spans beyond 4.5 m may be achieved by increasing the sectional
area of the rafters and ties. At little over 5 m the necessary size of
timber becomes uneconomical in comparison to introducing
additional members within the roof space. This is principally the
use of a third dimensional unit known as a purlin which runs
parallel to the wallplate and ridgeboard as detailed in Fig. 6.11.
The purlin provides immediate support to the rafters and is in turn
supported by struts, collars and hangers at every fourth rafter.
Struts depend on mid-span support from a load bearing partition,
so although overall spans of 7.5 m are readily achieved,
intermediate support cannot be omitted.

Triple roofs (trusses)

Triple or framed roofs were developed during the late 1940s and
early 1950s, as a rational approach to timber economy in roof
design. The principal components are prefabricated or

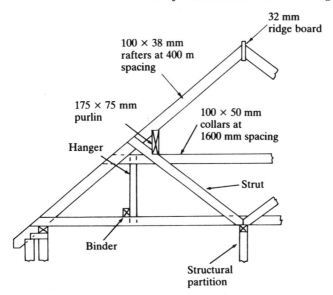

Fig. 6.11 Double roof purling detail

site-assembled trusses spaced at 1.8 m intervals to support purlins and ridgeboard. Intermediate construction is provided by three rafters and ceiling joists at 450 mm spacing. Varying designs offer an uninterrupted span potential of between 5 m and 11 m. Assembly is by simple bolted connections with double-sided toothed plates between adjacent timber sections. This is shown in Fig. 6.12 with the construction of a typical domestic roof truss of $22\frac{1}{2}°$ to 30° pitch, over 6 m span.

Trussed rafters

These are a series of triangular trusses which have gradually superseded the use of bolted trusses in domestic roofing. They originated in North America during the 1950s, but it was not until the early 1960s that they were marketed in the UK. They rapidly gained popularity offering the advantages of quality-controlled factory prefabrication with quick and simple site installation. Now, they are seen on virtually every housing site, as the expense of traditionally constructed craftsman-made *in situ* roofs is prohibitive, particularly on speculative developments.

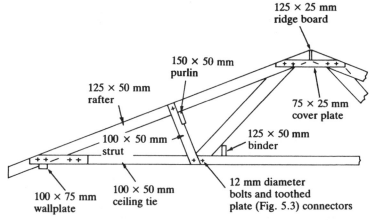

Fig. 6.12 Bolted truss

Timber must be structural quality, and planed on all faces to ensure accuracy and efficiency of location to galvanised steel nail plates. Minimum size is 35 × 95 mm (ex. 38 × 100) for spans up to 8.5 m and 45 × 95 mm (ex. 50 × 100) for spans between 8.5 m and 10 m. Precise span limits are difficult to define, as they depend on roof pitch, loading and arrangement of internal bracing. Most manufacturers offer a standard range of trusses to 12 m span with pitch variations between 15° and 35°. Purpose made designs are possible for larger spans and steeper pitches. Figure 6.13 shows some basic, popular truss patterns for modest span/loading requirements with symmetrical centre-line location of bracing relative to the spans and Fig. 6.14 butt-jointed timbers with nail plates either side to ensure continuity of load distribution.

Installation

Trusses spaced at maximum intervals of 600 mm, skew nailed to wallplates without the traditional birdsmouth notch detailed in Figs. 6.8 and 6.9. Nailing in this area is difficult because of the position of nailplates, therefore brackets of the type shown in Fig. 6.14 are preferred for a more positive fixing. In the absence of a ridgeboard and purlins, vertical and lateral stability are provided by:

(a) 38 × 25 mm tiling battens secured with 75 mm – 10 gauge galvanised steel nails;
(b) 100 × 25 mm longitudinal ties (Fig. 6.15);

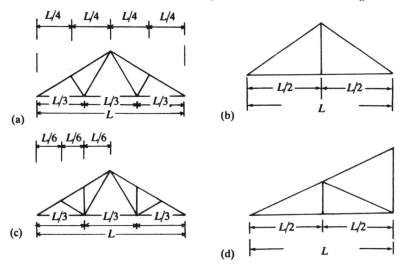

Fig. 6.13 Standard trussed rafters (a) Fink (b) Kingpost (c) Fan
         (d) Mono

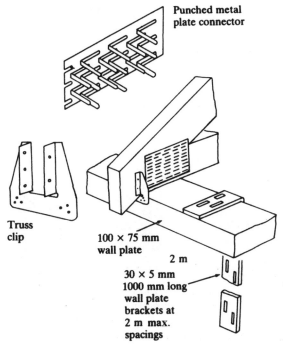

Fig. 6.14 Truss to wallplate detail

(c)   100 × 25 mm diagonal rafter (wind) bracing (Fig. 6.16);
(d)   gable ladders built into the gable wall (Fig. 6.17);
(e)   gable restraint straps (Fig. 6.18).

Note: Restraint straps are principally to provide lateral restraint to the gable wall from the roof structure, but in turn the wall provides complementary support to the roof.

Tiles and slate roof covering

Covering for pitched roofs to dwellings was for centuries limited to plain clay tiles, single lap clay tiles or slates. As roofing unit materials, both have a considerable history, but due to transport costs they are now restricted to the areas of production. Plain and

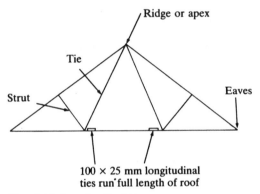

Fig. 6.15 Longitudinal ties

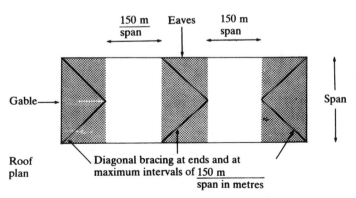

Located under rafter part of truss from eaves level to apex.
Fig. 6.16 Wind bracing

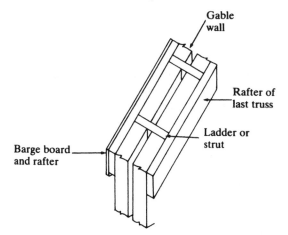

Fig. 6.17 Gable ladders

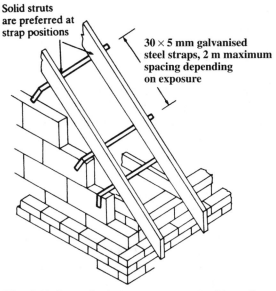

Fig. 6.18 Restraint between truss and gable wall

single lap tiles have also been produced from high quality concrete for some 70 years. In the last 20 years the desire to preserve slated roof tradition has led to the development of the asbestos-cement replica slate. These are produced to the same dimensions as slates, and fixing is very similar.

Plain clay or concrete tiles

These are also known as double lap tiles, as each tile overlaps two tiles, as may be seen in Figs 6.6 to 6.8. Also, each tile side laps half the two tiles below to prevent water penetrating the butted joints. Figure 6.19 shows a standard tile of 265 × 165 mm dimension with tile and a half and under ridge/eaves variations for gable verge courses and ridge/eaves courses respectively. Ridge tiles are produced in 300 mm or 450 mm lengths in half round, segmental or angular profiles for use on ridge or hips, where they are embedded in cement and sand mortar (1:3) with broken tile slips. Valleys and hips are shown in Fig. 6.20 with purpose-made valley tiles or tiles cut to a lead-lined valley. Hips are finished with ridge tiles terminated by a galvanised steel hip iron. Bonnet tiles are an alternative hip finish, with featured mortar joints.

Roof pitch should be at least 40° to ensure that driving rain cannot penetrate under or between the tiles, and nail fixing to battens is provided at four-course intervals. On pitches up to 55°, every third course is nailed. Above this angle every tile is nailed. Battens for tile fixing are 38 × 19 mm where rafter spacing is up to 450 mm, and 38 × 25 mm where rafters are between 450 and 600 mm. Spacing of battens is termed the gauge, which is dependent on the lap, i.e. the amount that each tile overlaps the tile but one below. This is normally 65 mm, but may increase to as much as 90 mm if in a situation of severe exposure. For example:

$$\text{Gauge} = \frac{\text{tile length} - \text{lap}}{2}$$
$$= \frac{265 - 65}{2}$$
$$= 100 \text{ mm.}$$

Battens are secured to rafters by galvanised steel or aluminium alloy nails with an underlay of bituminous felt draped over the rafters. This slightly improves the thermal insulation, and if laid parallel to the battens with 150 mm overlaps, provides a secondary waterproof layer should tiles become damaged or if snow is driven under the tiles.

Single lap tiles

These were originally produced from clay, in various shapes, sizes and formats. The Spanish or Italian (Roman) types are still manufactured and may be seen in use on many low pitch

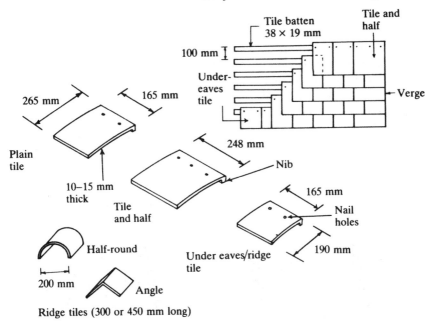

Fig. 6.19 Plain tiles

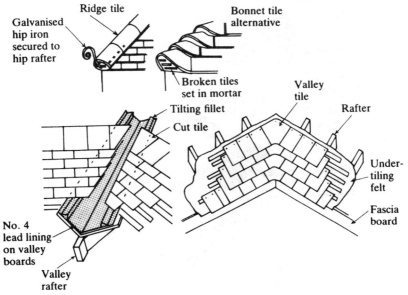

Fig. 6.20 Plain tiling to hips and valleys

continental roofs. In the UK, S-profiled handmade clay pantiles were a popular alternative to plain tiles, a tradition which has continued with machine made imitations from concrete. Size definition of single lap concrete tiles is difficult, as manufacturers patterns vary considerably. The examples shown in Fig. 6.21 are nominally 420 mm long × 330 mm wide, providing an actual exposed area of 345 mm × 295 mm assuming 75 mm head lap and 35 mm interlocking side lap. Pitches down to $17\frac{1}{2}°$ are possible with some patterns, and below $22\frac{1}{2}°$ a 100 mm head lap is advisable.

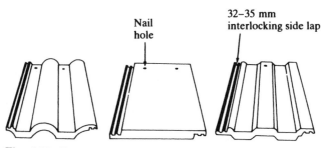

Fig. 6.21 Single lap concrete tiles

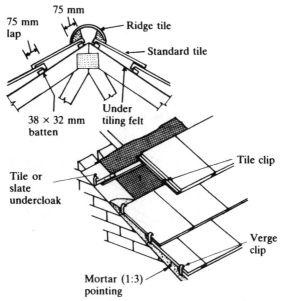

Fig. 6.22 Single lap tiled ridge and verge

Above 45° each tile is head nailed and if the pitch exceeds 55° clips are also necessary as shown in Fig. 6.22. Batten size is larger than for plain tiling; 38 × 32 mm for 450 mm rafter spacing and 50 × 32 mm for rafters between 450 and 600 mm.

$$\text{Gauge} = \text{tile length} - \text{lap}$$
$$420 - 75$$
$$345 \text{ mm}$$

Traditional slates

Slate is a naturally occurring, dense, impervious material, cut into fine layers to provide a small-scale roofing material capable of satisfying pitches down to 20°. Dimensions vary: the most popular, with their historic references, include:

Size (mm)	Name
400 × 200	Ladies
500 × 250	Countesses
600 × 300	Duchesses
600 × 350	Princesses
650 × 400	Empresses

Generally, the larger the slate the lower the roof pitch possible. However, the following table shows the relationship between pitch and lap:

Pitch (degrees)	Minimum lap (mm)
20	115
25	90
30–35	75
40–45	65

Nailing can be head or centre of the slate, through small holes pierced by the slater's zax (a hammer with a special point to the head). Centre nailing to the battens is preferable as the slate is less likely to be damaged by wind lifting. This method is as shown for composite slates in Fig. 6.23, but without the rivets. Batten gauging is also the same as composite slates and plain tiling, i.e. the slate length, less the lap divided by two, for example: Duchess slate at 30°,

$$\text{Gauge} = \frac{600 - 75}{2}$$

$$= 262 \text{ mm}$$

Asbestos slates

These slate replicas are produced to BS 690: Part 4 which specifies the following sizes:

Size (mm)	Lap (mm)	Gauge (mm)	Thickness (mm)
600 × 350	100	250	4
600 × 300	100	250	4
500 × 250	100	200	4
500 × 250	80	210	4

Note: Gauge is calculated with the same formula as for plain tiling,

$$\text{Gauge} = \frac{\text{slate length} - \text{lap}}{2}$$

Composition includes asbestos fibre, Portland cement and colouring pigments to provide dark blue, black and brown variations. The surface is coated with resin or acrylic silicon to improve the weathering qualities. Pitch range is between 20° minimum, and vertical for wall cladding. Fixing to battens is by centre nailing with two copper nails for each slate as shown in Fig. 6.23, and a copper disc rivet to prevent wind lifting damage. Each slate is tapped onto a rivet, which rests on the slate but one below. As the rivet penetrates the slate it is turned through 90° to provide a positive fixing. Ridge tiles are produced in half round or angular patterns with a 70 mm lap. They are secured with copper disc rivets penetration from the slates below. Alternatively, conventional ridge tiles may be used, bedded in cement and sand mortar.

Insulation and condensation

In the absence of insulation, approximately 35% of heat energy generated in a house is lost through the roof. Part L to the Building

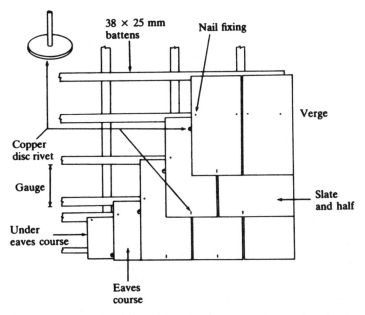

Fig. 6.23 Asbestos cement 'slates'

Regulations requires a roof structure to contain sufficient insulation to provide a maximum 'U' value of 0.25 W/m^2 K. The construction alone is seriously inadequate, hence the need for supplementary material between the rafters or ceiling joists. The latter is simpler, with insulation from either:

160 mm mineral fibre quilt
200 mm exfoliated vermiculite – loose fill
150 mm rigid mineral fibre slab
150 mm expanded polystyrene slab or loose fill.

High standards of insulation attract formation of condensation in the roof space, and subsequent potential decay to roof timbers. This must be averted with ventilation provided from a 10 mm continuous (or equivalent vents) gap along two opposing eaves finishes for roof pitches over 15°. Figures 6.8 and 6.9 show this provision, and Fig. 6.3 and 6.4 include the 25 mm gap required for roofs below 15° pitch. Additionally, high level ventilation equating to a continuous 5mm gap is required. This could be provided by ridge ventilator or eyebrow tiles as shown in Fig. 6.24.

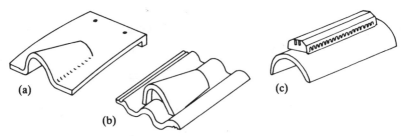

(a)

(b)

(c)

Fig. 6.24 Ventilation tiles (a) Plain concrete/clay eyebrow tile
(b) Interlocking concrete eyebrow tile (c) Ventilated
ridge tile

Wind effect on roof stability

Wind turbulence has a significant effect on the design of high rise
buildings and their immediate environment. Low-rise housing
design is less critical, although the effect of wind pressures (both
positive and negative) cannot be ignored. Figure 6.25 shows the
likely location of maximum wind forces around a roof, and the
map in Fig. 6.26 indicates division of the country into wind

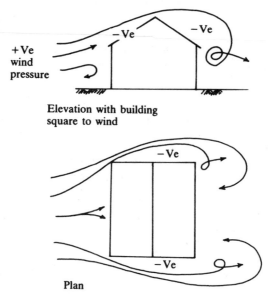

$-Ve$

$-Ve$

$+Ve$
wind
pressure

Elevation with building
square to wind

$-Ve$

$-Ve$

Plan

Fig. 6.25 Effect of wind loading

Wind speeds: A up to 44 m/s
                B up to 48 m/s
                C up to 52 m/s
                D over 52 m/s

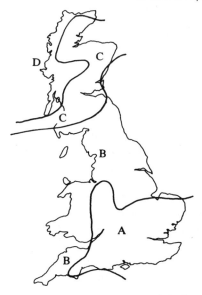

Fig. 6.26 Wind distribution (wind speeds: A up to 44 m/s, B up to 48 m/s, C up to 52 m/s, D over 52 m/s)

intensity zones which correspond with the following provision of bracketing between wall plate (see Fig. 6.14) and wall.

| Area | Exposure | | |
	Open country	Suburban	Town and city centres
A	1.6 m spacing	2.0 m spacing	2.0 m spacing
B	1.0 m spacing	1.25 m spacing	1.80 m spacing
C	1.0 m spacing	1.10 m spacing	1.50 m spacing
D	For all cases consult bracket manufacturers.		

7

Framed buildings

Framed buildings are constructed from beams and columns of reinforced concrete or steel sections. These form a framework or structural skeleton shown in outline in Fig. 7.1, into which non-load bearing partitions, external walls and floors are built or suspended.

The choice between reinforced concrete or steel will be determined by the building design, availability of materials and desired speed of construction. Whilst steel sections offer the advantages of immediate strength, simplicity of assembly and elimination of formwork, concrete provides greater design potential and does not require fire protection or maintenance.

Reinforced concrete cast *in situ*

Foundations

Column foundation bases are usually square on plan to distribute load evenly over two dimensions. Occasionally where the curtilage or obstructions intervene, the pad shape is rectangular. With both,

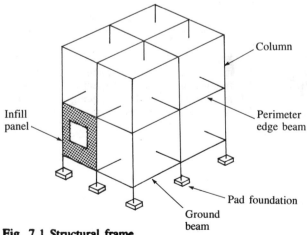

Fig. 7.1 Structural frame

the base area of concrete must be sufficient to prevent overstressing of the subsoil as shown by calculation in Chapter 3. Figure 7.2 shows a typical column pad foundation with reinforcement provided to resist both bending and shear stresses. A 75 mm kicker is necessary to locate column formwork accurately. Figure 7.2 also illustrates the construction of the kicker formwork which must be precisely manufactured and located, to ensure accurate column positioning.

In similarity with other foundations, column bases should be below ground level on suitable bearing strata. The pit excavation corresponds with the required shape and size of pad, therefore formwork is unnecessary. Occasionally, in poor subsoils, the column base bears on one or more reinforced concrete piles as shown in Fig. 7.3. Here, the base becomes a pile cap situated at ground level. Possible formwork is shown in Fig. 7.4.

Columns

Columns are most effective if designed with a uniform radius of

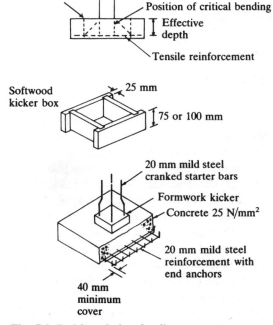

Fig. 7.2 Pad foundation detail

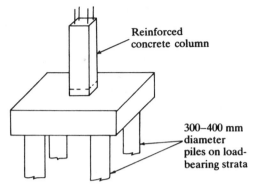

Reinforced
concrete column

300–400 mm
diameter
piles on load-
bearing strata

Fig. 7.3 Pile cap

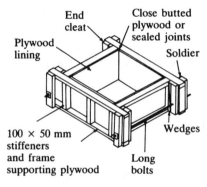

End
cleat

Close butted
plywood or
sealed joints

Plywood
lining

Soldier

100 × 50 mm
stiffeners
and frame
supporting plywood

Wedges

Long
bolts

Fig. 7.4 Pile cap formwork

gyration, i.e. round on plan. This shape is incompatible with other building units and costly to create, therefore most columns are designed and produced square on plan to create an equal radius of gyration in two dimensions (the x–x and y–y axes), and to permit comparatively simpler construction. Figure 7.5 shows uniform column shapes with a regular arrangement of longitudinal reinforcement to resist eccentric tensile stresses. The quantity should be between 1% and 8% of the column sectional area. In excess of 8%, it becomes difficult to compact and place concrete, therefore a steel I section would be a preferable substitute. Transverse reinforcement called links or binders are required to prevent buckling of longitudinal reinforcement. These are at least 6 mm diameter or a quarter of the longitudinal reinforcement (take greater value). Links are spaced at a maximum pitch of:

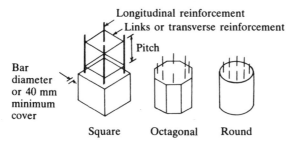

Fig. 7.5 Regular reinforced concrete columns

(a) the least lateral dimension of the column;
(b) 12 × smallest longitudinal reinforcement diameter;
(c) 300 mm (take least value).

Column formwork is from plywood, preferably shuttering grade, having one side only planed. It may be lined or faced with hardboard or a plastic sheeting to create a high-quality fair faced finish. Plywood should be pretreated with a release agent or mould oil to prevent cement grout adhering to the surface. Strength and stability are very important to withstand the hydrostatic pressure generated by placing of wet concrete. External bracing from timber or adjustable steel props is essential, and plywood rigidity is obtained from timber cleats or soldiers and steel clamps as shown in Fig. 7.6.

Beams

Reinforced concrete beams are larger examples of the design principles relating to concrete lintels explained in Chapter 4. In a simply supported situation, the areas of weakness will occur mid-span (maximum bending), and close to the support columns (maximum shear). Shear reinforcement could take the form of bars bent at 45° near to the supports, thus opposing the potential shear cracks at right angles. Alternatively, steel binders or links could be very closely spaced in this area. Figure 7.7 shows both possibilities with provision of shear reinforcement as a ratio of span.

Beam formwork is basically an open box, with the three sides cut to the required dimensions. Shuttering plywood provides the lining, with softwood struts and bracketing to enhance rigidity. Figure 7.8 shows this traditional approach to beam formwork with floor formwork included. Beam clamps are also shown; these offer

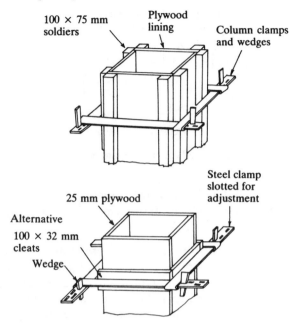

Fig. 7.6 Column formwork

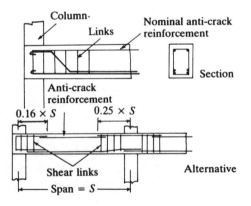

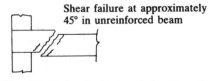

Fig. 7.7 Concrete beam reinforcement

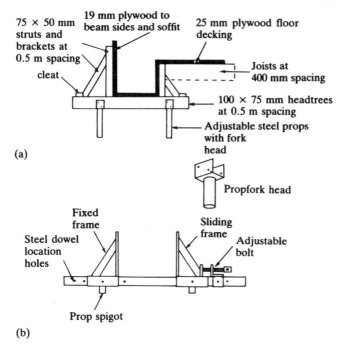

Fig. 7.8 Formwork to beams (a) Traditional (b) Adjustable beam
      clamps

a more progressive support system, having an adjustable facility to
accommodate various beam dimensions.

Brick infill panel walling

Non-load-bearing infill brick and block cavity walls are required to
provide an attractive finish to framed structures, adequate thermal
insulation, resistance to rain penetration and wind pressure while
still remaining sufficiently self-supporting and stable.

The stability and strength of the frame may be enhanced by the
infill wall, but for design purposes the frame is considered
independently, therefore settlement of the frame and differential
thermal movement must be absorbed between the two elements.
Possible constructional forms are shown in Fig. 7.9 for
single-storey steel portal-framed structures and Fig. 7.10 for
reinforced concrete or concrete clad steel-framed buildings.
Briquettes are shown as a possible non-structural surface
treatment to disguise the structural frame where the outer brick

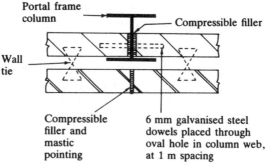

Fig. 7.9 Cavity infill wall with a steel column

wall is permitted to overhang the frame by not more than one third of the brick thickness.

Precast concrete

Pre-cast concrete framed buildings are a factory-produced alternative to casting concrete *in situ*. These have several advantages, in particular the precise control of unit production and exclusion from the effects of the weather. Structural improvement using prestressing techniques is also possible. On site concrete production plant, handling equipment, formwork and associated labour are unnecessary, saving considerable construction costs. Additionally, semi-skilled labour may be employed for the relatively simple bolted assembly. Design flexibility is restricted to the manufacturers' specifications for span, height, spacing of units, etc., but most systems incorporate sufficient variables to comply with the relatively regular structural form expected of this type of construction. The most common example of precasting the structural frame is in portal construction, although precast columns, beams and floor sections (Chapter 5) may be applied to multi-storey buildings.

Portal frames

These single-storey structures have widespread application for many building purposes. Their main advantage of clear unobstructed space from floor to rafters is compatible with the requirements for workshops, small factory assembly areas, warehousing and farm buildings. Frames are produced in two, three or four sections for spans ranging between 6 m and 25 m.

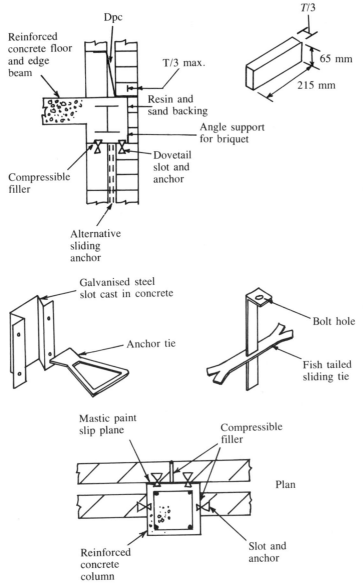

Dpc

Reinforced concrete floor and edge beam

T/3 max.

Resin and sand backing

Angle support for briquet

Dovetail slot and anchor

Compressible filler

Alternative sliding anchor

T/3

65 mm

215 mm

Galvanised steel slot cast in concrete

Anchor tie

Bolt hole

Fish tailed sliding tie

Mastic paint slip plane

Compressible filler

Plan

Reinforced concrete column

Slot and anchor

Fig. 7.10 Infill cavity panels to framed buildings

Three- and four-piece frames are also necessary for simplifying transport and site handling. Figure 7.11 shows a typical three-piece concrete portal for use in bay lengths of 4.5 m, 6 m or 9 m. Included with this drawing is a detail of the pocket foundation cast

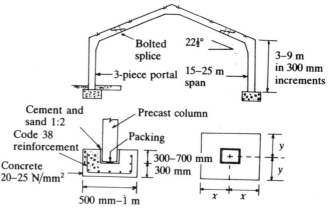

Foundation section and plan

Fig. 7.11 Reinforced concrete portal frame

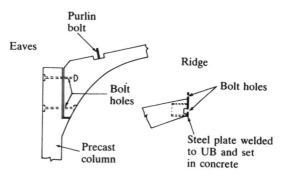

Fig. 7.12 Portal frame connection at eaves and ridge

in situ, necessary for this type of construction. The pocket, which often tapers, is formed with an expanded polystyrene mould or plywood box. Figure 7.12 shows an alternative, bolted eaves connection and the ridge connection for a four-piece frame.

Multi-storey construction

Precast components have successful application to buildings of up to four storeys. They are suitable for schools, offices and similar commercial buildings where a simple repetitive framework is acceptable. In order to vary appearance and to adapt to client

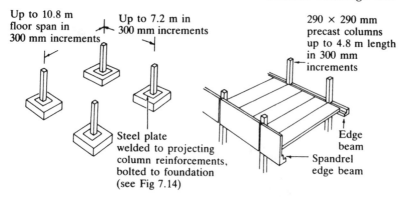

Up to 10.8 m
floor span in
300 mm increments

Up to 7.2 m in
300 mm increments

290 × 290 mm
precast columns
up to 4.8 m length
in 300 mm
increments

Steel plate
welded to projecting
column reinforcements,
bolted to foundation
(see Fig 7.14)

Edge
beam

Spandrel
edge beam

Fig. 7.13 Precast frame with columns and beams

requirements, manufacturers stock a large range of components
and accessories. Site assembly is again relatively easy as shown with
the foundation and floor elevations in Fig. 7.13, and the
construction details in Fig. 7.14. Ease of assembly must not be taken
for granted and close supervision is essential to ensure correct location
of components to avoid structural collapse, such as occurred in a gas
explosion at Ronan Point in the late 1960s.

Structural steelwork

Cast iron and wrought iron have a history of limited success as
materials for structural beams and columns. It was not until the
successful conversion of iron to steel between 1850 and 1870 that
steel became a viable alternative, although it took until 1897 to
convince the government that steel was capable of transmitting
greater stresses than wrought iron.

The first notable steel-framed building was erected in Chicago in
1885. In this country, the Ritz hotel in London was the first, built
between 1904 and 1905, shortly followed by a steel-frame
extension to Selfridges the following year. In contemporary
buildings, numerous examples exist in most towns and cities as
modern design principles permit favourable economic
comparisons with reinforced concrete. Steel is relatively light,
which allows large spans without intermediate support.
Foundation costs are also reduced, as the load transmitted to the
subsoil is less than that for equivalent strength masonry or
concrete.

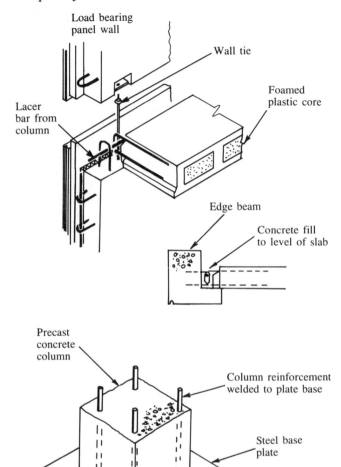

Fig. 7.14 Precast frame construction details

Steel sections

Various shaped profiles are produced by passing white-hot steel through a series of rollers which reduce the steel to the required size. During the early development of steel framed building the

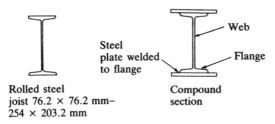

Rolled steel
joist 76.2 × 76.2 mm–
254 × 203.2 mm

Compound
section

Fig. 7.15 Steel joists

'I'-section girder known as a rolled steel joist was used for both beam and column applications. Figure 7.15 illustrates this section with the addition of steel plates welded to the flanges to produce a compound section, formerly used for columns. This still left a slender web which had very limited bearing capacity for direct bearing beams.

Universal sections

Demand for a more acceptable column section, plus the uneconomics of rolling tapered flanges, has led to the introduction of parallel flange sections with variations in web and flange thicknesses within the same serial size. These sections, produced to the dimensions specified in BS4, are known as universal beams

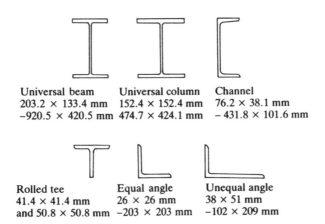

Universal beam
203.2 × 133.4 mm
–920.5 × 420.5 mm

Universal column
152.4 × 152.4 mm
474.7 × 424.1 mm

Channel
76.2 × 38.1 mm
– 431.8 × 101.6 mm

Rolled tee
41.4 × 41.4 mm
and 50.8 × 50.8 mm

Equal angle
26 × 26 mm
–203 × 203 mm

Unequal angle
38 × 51 mm
–102 × 209 mm

Fig. 7.16 Standard rolled sections

and universal columns. The principal differences are shown in Fig. 7.16. In general the column flange width is similar to the section depth, but this only applies to sections up to 305 mm dimensions. All universal beams are narrower across the flange than their depth and are manufactured in sizes which commence approximately where RSJs cease.

Open web beams

Open web beams are produced by castellating a standard RSJ or UB as shown in Fig. 7.17. The longitudinal oxy-acetylene cut divides the section into two, before it is welded back together in a different arrangement, to manufacture a section 50% deeper than the original. The purpose is to create a lightweight beam capable of modest loading over large spans with a high resistance to deflection. Additionally, the voids are very convenient openings for electrical conduits and piped services.

Lattice beams

An alternative method for producing lightweight open web beams involves fabrication from standard steel units. These are principally channels, box or angle sections, and dowels as an optional lacing. Assembly is by welding, a technique compatible with the range of sectional units which may be combined to provide considerable design flexibility. Figure 7.18 shows several

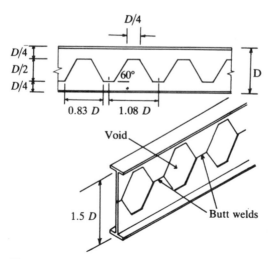

Fig. 7.17 Castellated beam

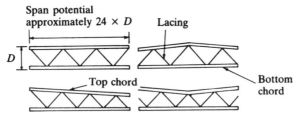

Fig. 7.18 Lattice beams

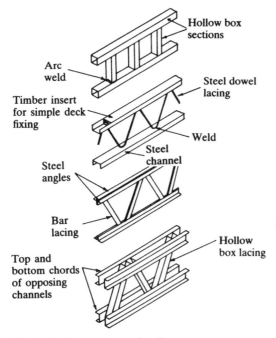

Fig. 7.19 Lattice beam details

standard profiles for spans up to 15 m, with some unit details in Fig. 7.19. Spans in excess of 15 m are possible, but are unlikely to be economic because of the cost of design, the size of component required and the procedure for non-standard assembly.

Cold-rolled sections

Cold-rolled steel sections are gaining in popularity over the traditional use of white-hot rolled universal sections, RSJs etc. They are more effective in the use of steel, allowing the structural engineer to select a profile which requires comparably less steel for the given design load. The production price per tonne is higher, but economy in design permits significant saving in weight. These sections have enjoyed considerable use in school building programmes where rational use of architecture has provided a repetitive form of construction. This was known as CLASP (Consortium of Local Authorities Special Programme) and it originated when Nottingham Council combined with other local authorities to create a scheme from which they could all benefit by mass and repetitive production. The system has since been adapted to offices, flats and commercial developments up to four storeys high. Figure 7.20 shows some of the most common cold-rolled section used by CLASP and subsequent systems.

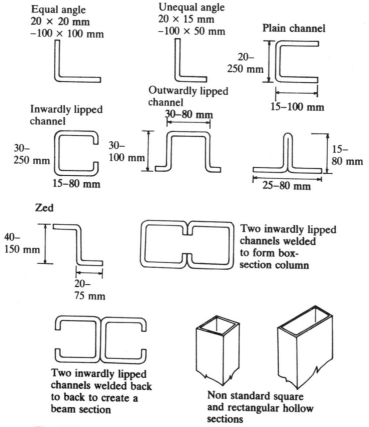

Fig. 7.20 Standard cold-rolled steel sections

Cold-rolled purlins

Steel trusses and portal frames have traditionally used hot-rolled steel angles as purlin support to corrugated decking. These are gradually being superseded by ZED and SIGMA sections. This is because:

(a) their cost compares favourably;
(b) sheet fastenings may be effectively shorter, as they are simple to locate around the upper flange lip;
(c) their shear centre is contained within the section, which makes them more resistant to twisting than angles or channels.

Figure 7.21 shows both sections.

Column foundation

The load transmitted by a steel column is very high relative to the material cross-sectional area. Therefore it is essential to spread this load over sufficient bearing area to prevent a shearing effect. Figure 7.22 shows provision of a steel base plate for this purpose, with the option of angle cleats and gusset plates where loads are exceptional. Below the base plate, a square pad of concrete is required to transmit the column load to surrounding subsoil. This pad is reinforced to resist shear and bending as detailed in Figs 7.2 and 7.31 with the addition of projecting bolts to locate and position each column accurately. Accuracy is ensured by positioning the bolts in the wet concrete pad with a bolt 'box' or template. This is shown in Fig. 7.23 composed of a plywood or chipboard sheet with holes and bolts secured to match the holes in the column base. An expanded polystyrene sleeve and a large square washer are attached to each bolt. The plywood and polystyrene are removed after the bolts are securely fixed to the

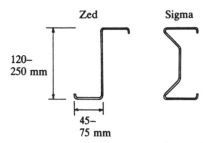

Fig. 7.21 Zed and sigma section purlins

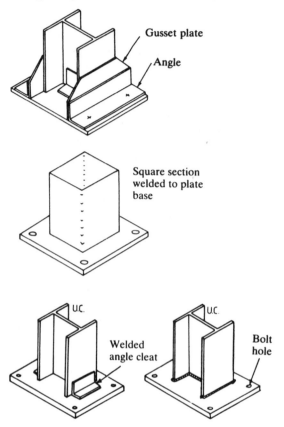

Gusset plate

Angle

Square section
welded to plate
base

U.C.

U.C.

Welded
angle cleat

Bolt
hole

Fig. 7.22 Steel column bases

set concrete, leaving a gap around each bolt which permits slight
movement to absorb any minor inaccuracy as the column is
lowered onto the bolts. Steel packing under the column allows
adjustment with the bolts until the column is vertical. The
remaining space below the column base and around each bolt is
packed with a fairly dry strong cement and sand mortar of 1:1
ratio. This is known as a rigid or fixed base.

Fastenings

Fabrication and assembly of a steel framework is by electric arc
welding or bolts. Welding is virtually restricted to factory
prefabrication, as the inconvenience on site compares un-
favourably with the simplicity of bolted connections.

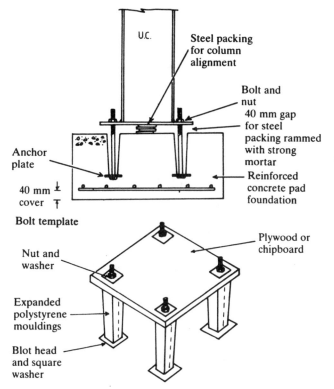

Fig. 7.23 Fixed column base detail

Welding

Welded connections are produced by an electric arc process which uses an expendable steel rod as an electode. It is simple and quick, but requires accurate location of components, a factor difficult to achieve on site, hence the preference here for bolts in preformed holes. Butt welds are more efficient for direct transfer of load, although less practical for joining adjacent sections. Fillet welds as detailed in Fig. 7.24 are more convenient.

Bolts

For structural engineering applications, three types of bolt are applicable:

(a) black bolts;
(b) machined or turned and fitted bolts;
(c) high-strength friction-grip bolts.

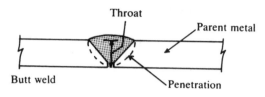

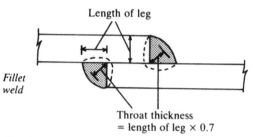

Fig. 7.24 Types of weld (a) Buttweld (b) fillet weld

(a) Black bolts. These are the cheapest of the three categories, manufactured by drop forging and machining of the threads. Dimensional accuracy is poor and a hole clearance of about 2 mm is expected. However, they are adequate for most direct bearing situations, where bolt shear is minimal.

(b) Machined bolts. These are an improvement on black bolts, with a fully machined specification to provide precise tolerance. Clearance between bolt and hole is within 0.5 mm and it is often necessary to hammer in these bolts because of the excellence of fit. This efficiency allows greater strength transfer, but there is no margin for dimensional adjustment in the frame.

(c) High-strength friction-grip bolts. These are produced from high-tensile steel to withstand greater stress loading than the other bolts. Their effectiveness is achieved by tightening to a predetermined shank tension or torque so that the clamping force transfers the load by friction between adjacent sections and not by shear or direct bearing on the bolts. These are more expensive than the other bolts, but the quantity necessary for each connection is less.

Standard connections

The principal connection in a steel-framed building between

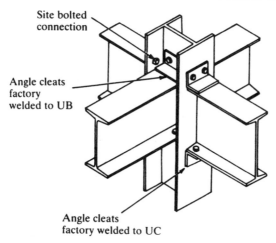

Site bolted connection

Angle cleats factory welded to UB

Angle cleats factory welded to UC

Secondary beam notched

Elevation on A

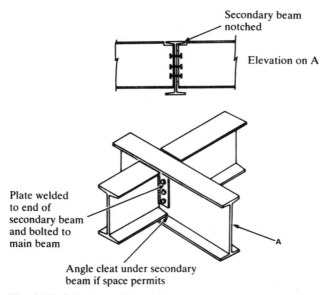

Plate welded to end of secondary beam and bolted to main beam

A

Angle cleat under secondary beam if space permits

Fig. 7.25 Column and beam connections

column and beam is shown in Fig. 7.25 with beam connections to the web and flange of a UC. Angle seating cleats are welded to the column to allow direct bearing from the beam. This minimises any shear transfer in the bolts, which function merely as a means of locating members. The top angle cleat is usually smaller, it helps to prevent the beam twisting and bending under load. For

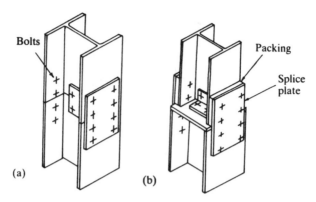

Fig. 7.26 column splices (a) Plain splice (b) Reducing splice

convenience of site assembly this cleat is welded to the beam and bolted to the column. The other connection shown in Fig. 7.25 is between a main beam and a secondary beam. Here, the less significant beam is notched in the top flange to preserve uniform top surface levels and a plate is welded to the web for direct bolting to the web of the main beam. Figure 7.26 shows a splice connection between two columns of equal size. Also shown is the more likely reducing splice, which occurs because of the reduced column load as the frame increases in height. Horizontal splices should not occur, beams should span uninterrupted between columns or other supplementary support.

Roof trusses

Steel trusses are a lattice frame composed of opposing rafters, internal bracing and a ceiling tie. The arrangement of bracing will determine whether the members are in compression (strut) or tension (tie). Figure 7.27 shows several truss patterns for spans up to 30 m, using standard angle sections and plate gussets at intersections. It is important that centre lines of angles coincide for concentric distribution of load. This is shown in Fig. 7.28 with a detail of a truss capable of 10 m span. Purlins may be steel angles or zed sections bolted to angle cleats. Alternatively sigma sections could be used with specially made cleats (see Fig. 7.32). Some other purlin applications are shown in Fig. 7.29 with bolt or screw fixings for the decking material. Decking is usually corrugated asbestos, profiled galvanised steel, aluminium or clear perspex. Spacing of purlins are arranged to suit decking materials with

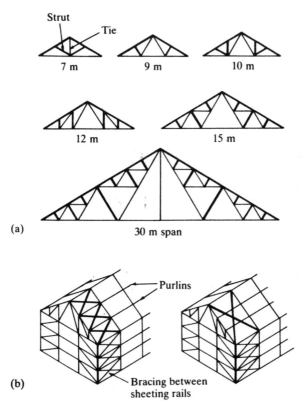

(a)

(b)

Fig. 7.27 (a) Steel trusses (b) Wind bracing alternatives

minimum overlaps to suit roof pitch. Ridge and eaves treatment is shown in Fig. 7.30.

Pitch	Minimum lap
$10° - 15°$	150 mm, mastic sealed
$15° - 22\frac{1}{2}°$	150 mm, mastic sealed or
	300 mm end lap only
Over $22\frac{1}{2}°$	150 mm end lap only

Portal frames

Steel portals are continuous plane frames manufactured from universal beams and columns. Height and span are limited only by

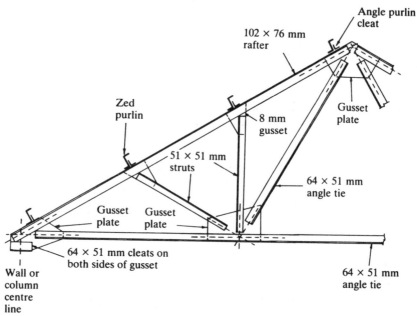

Fig. 7.28 Steel truss detail

material economies, which restrict most applications to a single storey and a span of 30 m. The principal advantage of this type of frame is provision of a large clear floor area with space from floor to ridge. This offers adaptability for many purposes, particularly assembly, manufacture and storage, an essential feature of speculative developments where the end user is unknown. Rafters have a preferred pitch of between 5° and 15°, although up to 22½° is possible. Bay spacing is at 4.5, 6, 7.5 or 9 m intervals to correspond with soil conditions, frame loading and cladding technique. Figure 7.31 shows typical details of a steel portal, including the use of gussets at eaves and ridge for enhanced rigidity.

Wind bracing and anti-sag bars

Wind effect will be most significant in the end bays, so to resist the negative and positive pressures in this area standard angle, channel or box sections are diagonally braced as shown in Fig. 7.32. Also shown is an application in exposed situations, where the purlins will require stabilising with restraint or anti-sag bars. These are threaded steel dowels bolted between positions

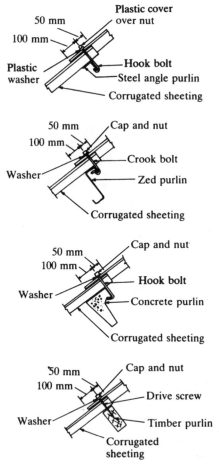

Fig. 7.29 Purlin details

near the top flange of one purlin and the lower flange of the adjacent purlin.

Fire protection

Although steel is incombustible, its performance in fire is poor. Under load, a section attaining a temperature between 500°C and 550°C loses strength, will probably buckle and no longer support its superimposed structure. Consequently, Part B to the Building Regulations imposes severe limitations on the use of unprotected steelwork, and provides detailed schedules of materials which will satisfy the mandatory period of fire resistance.

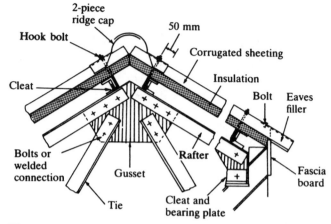

Fig. 7.30 Steel truss, ridge and eaves

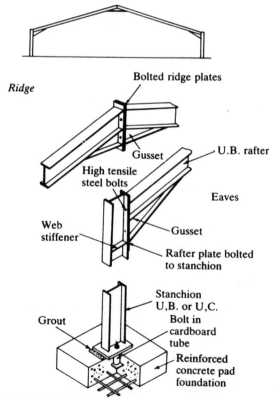

Fig. 7.31 Portal frames

Plan of portal frames

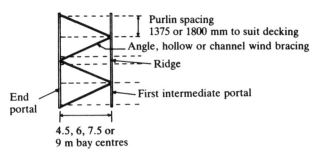

Purlin spacing
1375 or 1800 mm to suit decking
Angle, hollow or channel wind bracing
Ridge
End portal
First intermediate portal
4.5, 6, 7.5 or
9 m bay centres

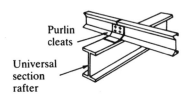

Purlin cleats

Universal section rafter

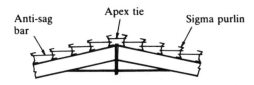

Anti-sag bar

Apex tie

Sigma purlin

Fig. 7.32 Purlin fixing and bracing

The application may be solid, i.e. a fully enclosed section, or hollow, where central voids occur as a result of encasing the unit. Examples of both are shown in Figs 7.33 and 7.34 respectively. Examples of *solid* fire protection

- Concrete
- Bricks
- Blocks of lightweight or dense concrete
- Cement and asbestos, sprayed or trowelled on
- Cement and vermiculite, sprayed or trowelled on
- Intumescent paint

Examples of *hollow* fire protection

- Bricks
- Blocks of lightweight or dense concrete
- Expanded metal lathing with rendered finish, e.g. metal

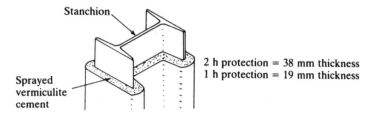

2 h protection = 38 mm thickness
1 h protection = 19 mm thickness

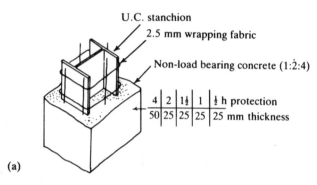

(a)

Fig. 7.33 Fire protection (a) Solid (b) Hollow

lathing plaster, vermiculite or perlite-gypsum plaster cement and asbestos or vermiculite
- Plasterboard
- Asbestos boarding
- Proprietary slabs of acceptable composition and thicknesses.

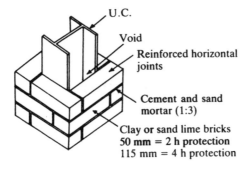

U.C.

Void

Reinforced horizontal joints

Cement and sand mortar (1:3)

Clay or sand lime bricks
50 mm = 2 h protection
115 mm = 4 h protection

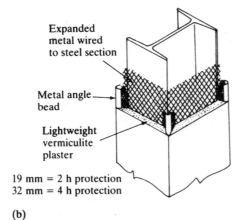

Expanded metal wired to steel section

Metal angle bead

Lightweight vermiculite plaster

19 mm = 2 h protection
32 mm = 4 h protection

(b)

Fig. 7.33 Cont'd

8

Fireplaces and flues

A log- or coal-fuelled open fire has been the central heat source in homes since the earliest evidence of man's existence. Its brief demise occurred during the 1960s, when 'clean' heating from radiators became financially viable in most homes. Fireplaces were considered dirty and old-fashioned, and many were bricked up and decorated over. The 1970s and 1980s have seen a revival in the use of solid-fuelled fires. This has resulted from escalating central heating fuel costs and the design of many attractive fire surrounds and log-burning devices which enhance living areas and provide a central focal point.

Fire recess

The recess for a fireplace is constructed by attaching piers to the wall which project into the living area, or by offsetting the cavity brickwork into the space outside. The latter is preferred as it saves floor space, but heat loss from the flue is increased. In either case the brickwork is supported by concrete projections from the foundations. These are shown in Fig. 3.8 and explained in the corresponding text.

The structural dimensions around the hearth are detailed in Part J to the Building Regulations. Most are shown in Fig. 8.1 which includes various fire recess arrangements with the minimum jamb and back thickness of 200 mm. The back may reduce to 100 mm if it serves two recesses built on opposite sides of a partition, and also if it functions as the inner leaf of an external wall. In all cases, the dimensions given must contain bricks or blocks of clay or concrete or concrete cast *in situ*.

Hearth

Hearths are constructed of concrete to a minimum thickness of

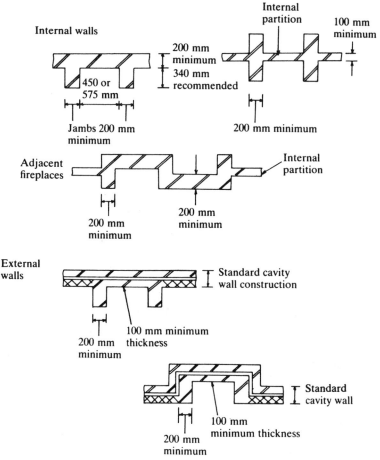

Fig. 8.1 Fireplace formations

125 mm. The concrete must extend fully into the fire recess and project at least 500 mm in front of the jambs and 150 mm on each side of the opening. Figure 8.2 illustrates these requirements and includes the associated construction around a typical suspended timber ground floor.

Construction in the fire recess

Fireplace openings, in similarity with other wall openings, require a lintel to bridge the space immediately above the fire and to support the superimposed brickwork. Also a fire back is necessary to

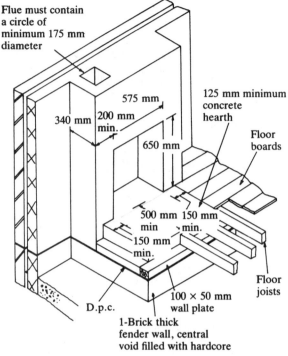

Fig. 8.2 Fireplace and timber ground floor

deflect most of the heat forward, otherwise this is lost up the chimney.

Lintels and throat units

Concrete fireplace lintels are triangular in cross section to avoid obstructing the flow of smoke and combustion gases. Each end is rectangular for building into brickwork and a single mild-steel rod spans from each end to provide resistance to tensile loads. Figure 8.3 shows the details, and Fig. 8.6 a section through the fire recess to indicate the installation. The brickwork each side is corbelled and rendered smooth with a cement, lime and sand (1:2:9) parging. Alternatively, a throat unit may be used as a more efficient means of connecting fire to flue. These are made from heat-resisting fire clay or concrete in various shapes. Two examples are shown in Fig. 8.4, with detachable front plates to allow accessibility during construction. Figure 8.5 shows installation.

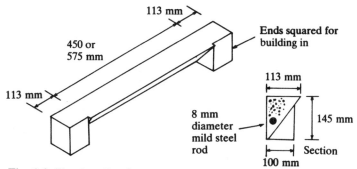

Fig. 8.3 Fireplace lintel

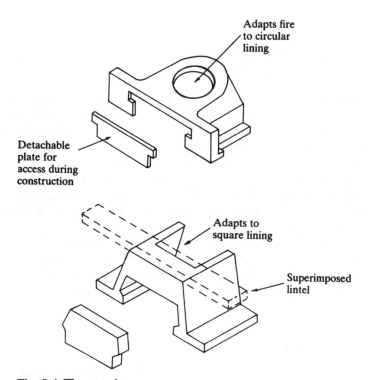

Fig. 8.4 Throat units

Fireback

A fireback of heat-resistant fire clay is built into the rear of the fire opening and positioned with its front edge aligned with the face of the jambs. This leaves a void behind the unit which is filled with an

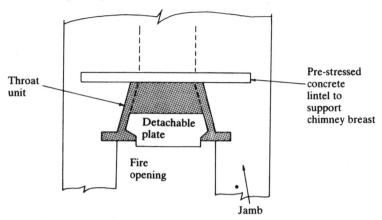

Fig. 8.5 Throat unit installed (exposed brickwork, lintel and throat
unit covered by attached fireplace surround or feature
stone or brickwork)

insulative concrete of vermiculite and cement or clean brick rubble
and mortar. Figure 8.6 shows the installation, and includes the
important layer of cardboard, placed to prevent adhesion between
backing material and fire back. A bond between these would
restrict natural expansion and contraction by the fire back and
cause cracking. The unit itself is usually produced in one piece, but
may be obtained in two, four or six pieces for asbestos tape and
boiler cemented joints to reduce the possibility of movement
cracks.

Firebacks of cast iron plate are also possible. These are
attractive in fully bricked fireplaces, particularly if embossed with
ornate designs. They help to reflect heat forward and prevent
damage to the rear wall, but lack a projecting knee to deflect heat
into the room.

Flue and lining

A flue serving an open fire must contain a clear circular space of at
least 175 mm diameter. Throughout its length offsets and bends
should be minimal and the angle of travel must exceed 45°. Flues
must be lined to prevent the products of combustion (sulphur, tar,
acid and ammonia) decomposing the bricks and mortar joints.
Many chimneys are parged on the inside with a trowelled cement
solution, but this is of limited success. Current practice is to
employ one of the following:

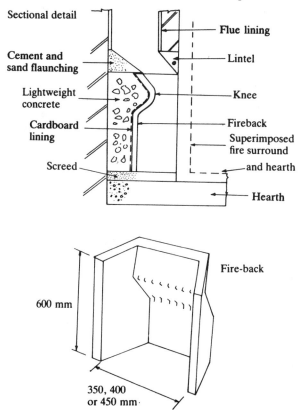

Fig. 8.6 Fireplace recess and fireback

(a) rebated and socketed clay linings to BS1181;
(b) rebated and socketed kiln-burnt aggregate and high alumina cement linings;
(c) spigot and socket drain pipes to BS65.

All categories are built in with socket uppermost and jointed with cement mortar.

An alternative is concrete flue blocks of kiln-burnt aggregate and high alumina cement. They could be used for construction of the chimney, thereby incorporating a lining within the structure. Flexible stainless steel linings are not recommended as they corrode and disintegrate in the presence of gases and condensation produced by solid fuels. Acceptable flue linings are shown in Fig. 8.7.

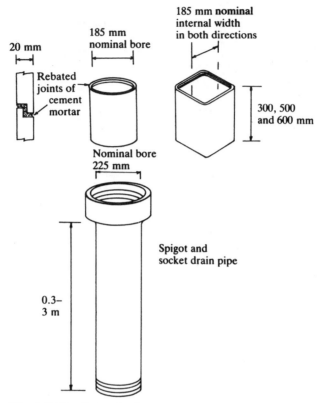

Fig. 8.7 Flue linings in clayware

Outlets for flues

In the interests of preventing fire the outlet to a flue and chimney must be carried to sufficient height above the roof. In all forms of construction quoted, the dimensions given include the brick structure only, excluding the pot or any attached flue terminals.

Flat roofs, i.e. of pitch less than 10° – 1 m minimum height above the finish.

Pitched roofs, projection at least 600 mm above the ridge capping. Elsewhere other than the ridge, the chimney extends at least 1 m from its highest intersection with the roof, unless it is within 600 mm horizontal distance of the ridge then the height may reduce to 600 mm above the intersection. Finally the proximity of ventilation inlets or outlets, opening windows and flue outlets must also receive consideration. This particularly

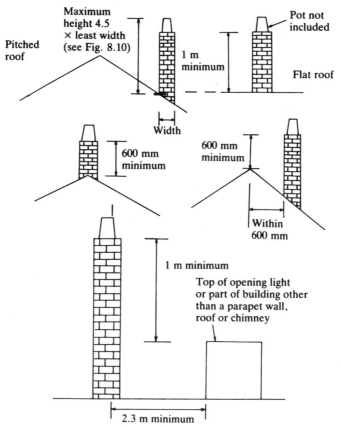

Fig. 8.8 Flue outlets

applies to upper floor windows, dormer windows and rooflights. Here, the chimney structure is at least 1 m above the opening light, and the flue outlet at least 2.3 m measured horizontally. The same requirements apply to any other part of the building other than the roof, a parapet or another chimney, as may be seen in Fig. 8.8 with the other considerations affecting flue outlets.

Chimney construction

Chimney design ranges from elaborate examples of feature brickwork, still much in evidence on Tudor buildings, to the modern plain brickwork with simple flaunching or precast concrete coping shown in Fig. 8.9. Experimentation with shapes, varying heights, offsets and terminals has provided some unusual

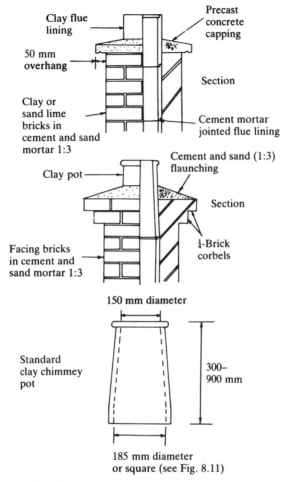

Fig. 8.9 Flue terminals

chimney presentation in many of our historic buildings. Chimney height is often extreme, presumably in an attempt to create an effective draw on the fire but with little regard for slenderness and structural stability. Modern requirements impose a minimum effective flue height of 4 m from the fire recess lintel or throat unit soffit, to the top of the flue terminal, in addition to the minimum chimney height dimensions shown in Fig. 8.8. Furthermore, the structural stability legislation in Part A to the Building Regulations imposes a maximum height dimension of $4\frac{1}{2}$ times the least lateral dimension. This is effected from the point that the chimney

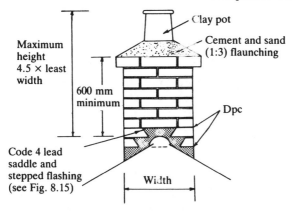

Maximum height 4.5 × least width

600 mm minimum

Clay pot

Cement and sand (1:3) flaunching

Dpc

Code 4 lead saddle and stepped flashing (see Fig. 8.15)

Width

Fig. 8.10 Chimney projection at ridge

intersects with the roof and the least width dimension of the chimney, as may be seen in Fig. 8.10.

Proximity of structural timber

Where the chimney passes through the upper floor, ceiling and roof framework it is important to fully isolate the timber. The minimum clear space between chimney face and adjacent timber

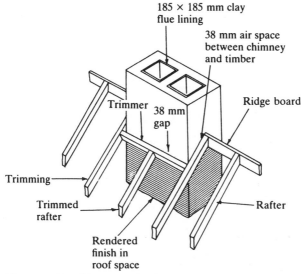

185 × 185 mm clay flue lining

38 mm air space between chimney and timber

Ridge board

Trimmer

38 mm gap

Trimming

Trimmed rafter

Rendered finish in roof space

Rafter

Fig. 8.11 Roof framework trimming to chimney

frame is 38 mm unless the chimney wall is at least 200 mm thick. Arrangement of joists and rafters is similar to stair openings (see Fig. 5.4) with the use of trimmer, trimming and trimmed members.

Figure 8.11 shows the treatment of rafters in a traditional roof around a ridge penetrating chimney. The trimmer in this example functions as support to the trimmed rafters and as continuity in place of the severed ridgeboard.

Weathering

An abutment between the side or cheek of a chimney and the tiling will require weathering with a stepped side flashing and soakers. If the chimney penetrates the roof between ridge and eaves a front apron and back gutter are also necessary to prevent water and snow penetration. Eaves chimneys require stepped side flashings, soakers and back gutter only and ridge chimneys require stepped side flashings, soakers, aprons and a ridge saddle each side. In every case, two damp-proof courses should be installed to coincide with the back gutter or saddle and front apron as shown in Fig. 8.12. Sheet materials suitable include copper, aluminium, zinc and nuralite, but lead by virtue of its practicality and malleability is the most popular. Code 4 (1.8 mm) is adequate for most situations, and the slightly thicker Code 5 (2.24 mm) reserved for conditions of severe exposure or for long lengths.

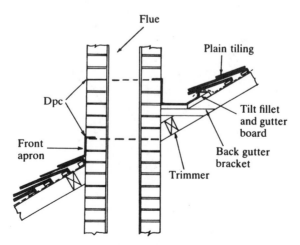

Fig. 8.12 Section through chimney

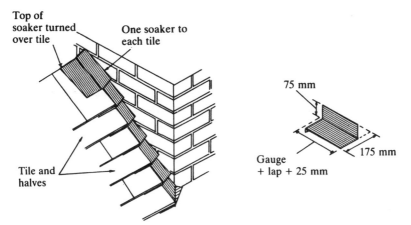

Top of
soaker turned
over tile

One soaker to
each tile

Tile and
halves

75 mm

175 mm

Gauge
+ lap + 25 mm

Fig. 8.13 Chimney abutment – soakers

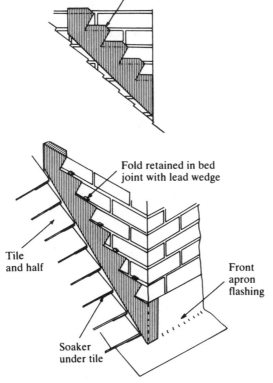

Folds coincide
with top edge of brick

Fold retained in bed
joint with lead wedge

Tile
and half

Front
apron
flashing

Soaker
under tile

Fig. 8.14 Chimney abutment – flashing

Where the roof covering is corrugated or highly profiled, typical of most modern single lap tiles, soakers are unnecessary as the side flashing is attached to the wall and dressed down onto sufficient area of roof covering. Otherwise, as with plain and flat concrete tiles and slates, soakers are cut, bent and placed over every tile as shown in Fig. 8.13. The stepped side flashing is placed against the chimney side as shown in Fig. 8.14 and cut in levels to follow the brick courses, with an allowance of 25 mm for wedging and mortaring into the bed joints. Front apron, back gutter and ridge

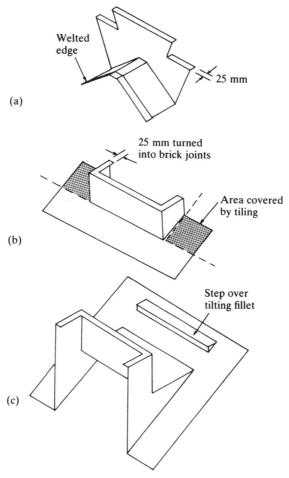

Fig. 8.15 Chimney flashing extras (a) Saddle (b) Apron (c) Back gutter

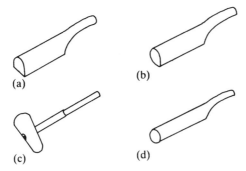

Fig. 8.16 Lead sheet tools (a) Dresser (b) Setting in stick
(c) Mallet (d) Bossing stick

saddle are shown in Fig. 8.15. These are made by cutting and welding the lead to suit the chimney size and roof pitch. Alternatively they may be manipulated into shape using the traditional plumbers' wooden bossing tools shown in Fig. 8.16.

9

Doors and windows

Doors

The fundamental purpose of a door is to provide access into or out of a building, and between the various compartments within a building. Additionally, the following functions are fulfilled, the extent depending on the building type and purpose:

1 Security.
2 Weather resistance.
3 Fire resistance.
4 Thermal insulation.
5 Sound insulation.

Frames and linings

Openings in walls are considered in Chapter 4, where provision of lintels and other structural factors are illustrated in the appropriate section. The method of securing a door to the wall will be determined by the door position. External doors are hung in frames which incorporate various features to prevent draughts and damp penetration, whilst internal doors are located in linings of much simpler construction.

Frames

Timber door frames are constructed from a sill, two posts or jambs and a head. The sill, being exposed, should be manufactured from hardwood to contain a weathered front edge, a drip and mortar groove to the underside and a recess for a metal or plastic weather bar on the top. Posts and head are usually softwood. They are the same section, containing a rebate for the door to include an anti-capillarity groove, a splayed edge and a mortar groove. Joining of the frame members is by mortice and tenon joints with round hardwood dowels or metal star shaped dowels as shown in

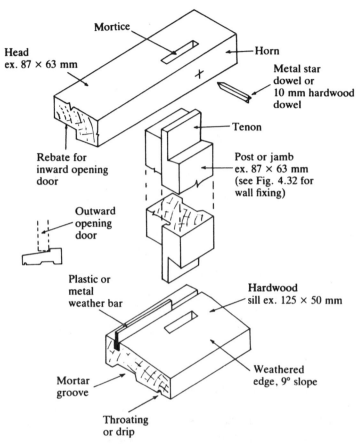

Head
ex. 87 × 63 mm

Mortice

Horn

Metal star
dowel or
10 mm hardwood
dowel

Tenon

Rebate for
inward opening
door

Post or jamb
ex. 87 × 63 mm
(see Fig. 4.32 for
wall fixing)

Outward
opening
door

Plastic or
metal
weather bar

Hardwood
sill ex. 125 × 50 mm

Mortar
groove

Weathered
edge, 9° slope

Throating
or drip

Fig. 9.1 Door frame details

Fig. 9.1. Also shown is the sill section used where a small entrance lobby or kitchen necessitate an outward opening door. In this situation, the posts and head are rebated to face externally.

Linings

Door linings are used as a means of trimming internal walls at junctions with doors. They consist of two jamb linings rebated into a grooved head lining as shown in Fig. 9.2. Plain linings are moulded from one piece of timber, with a rebate similar to a door frame to act as a door stop. The alternative, shown attached to a timber stud partition in Fig. 4.25, uses a planted stop nailed to a 32 mm or 25 mm thick planed standard section wide enough to

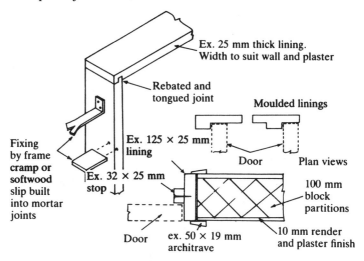

Ex. 25 mm thick lining.
Width to suit wall and plaster

Rebated and
tongued joint

Moulded linings

Fixing
by frame
cramp or
softwood
slip built
into mortar
joints

Ex. 125 × 25 mm
lining

Ex. 32 × 25 mm
stop

Door   Plan views

100 mm
block
partitions

Door   ex. 50 × 19 mm
architrave

10 mm render
and plaster finish

Metal linings (11.25 or 1.6 mm mild steel sheet)

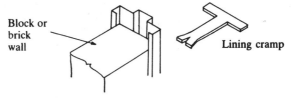

Block or
brick
wall

Lining cramp

Fig. 9.2 Door lining details

include the wall thickness and plaster finish. Occasionally metal
linings are specified where a high degree of fire resistance is
required, or where the doorway is likely to be subject to
considerable abuse. These and other door lining applications are
illustrated in Fig. 9.2.

Door types

Timber doors may be grouped into four separate categories,
determined by the mode of manufacture:

(a) Matchboarded or battened
(b) Panelled
(c) Flush
(d) Firecheck, flush

Matchboarded or battened doors (see Fig. 9.3)

The simplest matchboarded door consists of chamfered edge

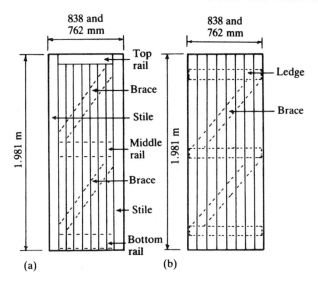

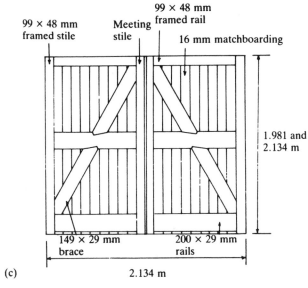

Fig. 9.3 Matchboarded doors (a) Framed, ledged and braced door
(b) Ledged and braced door (c) Garage doors

tongued and grooved battens or boards held vertical by three
horizontal ledges. The upper and lower ledge carry the 'T' hinges,
and the middle ledge provides a fixing for a latch or lock. With use
this door has a tendency to sag at the closing edge, as the only

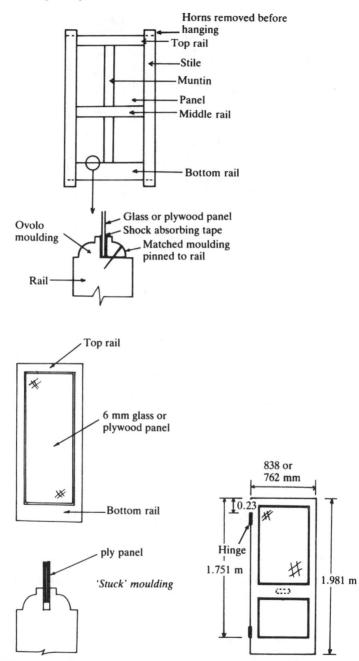

Fig. 9.4 Panelled doors

resistance is from the nails between ledges and battens. A slight improvement is the use of braces which incline from the hanging edge to resist the sagging effect. These are not always sufficient to resist warping, a characteristic noticed when a door fails to seat fully into the frame rebate.

The most satisfactory arrangement includes an outer framework stiffened with ledges (now termed rails) and braces. This robust and very serviceable unit is an excellent exterior door for garages and similar outbuildings, whereas the other two simpler versions are limited to well-sheltered exterior locations or interior use in cottage or farmhouse type buildings where they are in keeping with other traditional features.

Panelled doors (see Fig. 9.4)

These doors consist of a timber outer frame with intermediate dividing bars known as muntins and rails, and infill panels of glass or plywood. The style and quality vary considerably from simple single-panelled doors to purpose-made hardwood reproduction period doors with specially moulded panel work. Some examples of these are shown in Fig. 9.5. Joints between panel and frame may be square edge, chamfered or more elaborately moulded. The style and method used will depend on the door quality. The practice with most doors and glazed panels is to leave a recess for the panel, which is finished with a planted moulding as shown in Fig. 9.4. Better quality doors have mouldings 'stuck' or worked out of the solid. Joints between frame members are made with dowels as detailed in Fig. 9.6 or the more traditional mortice and tenon with wedges.

Flush doors (see Fig. 9.7)

Flush doors are very popular for interior use, as the smooth plain surface is easy to clean and decorate. The appearance can be chosen from a huge range of decorative veneers, painted hardboard, plastic or plywood to suit most building functions. They may also have external use with exterior grade plywood facing, but the plain finish is not so attractive in this situation. The construction varies, depending on the purpose. The simplest has a rectangular frame with a light timber skeleton core within. Alternative core materials include glued wood shavings or egg trays. Lock and bolt blocks are glued in place if required. Flaxboard panels are an alternative filling to provide a semi-solid composition with fire-resisting properties, and a solid core of

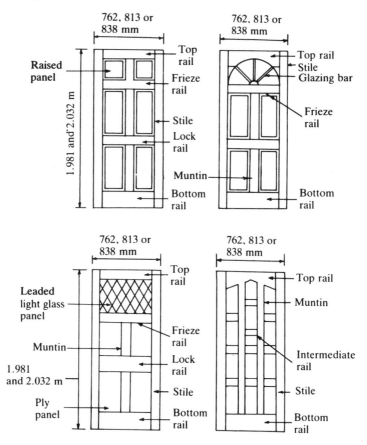

Fig. 9.5 Reproduction and purpose-made panel doors

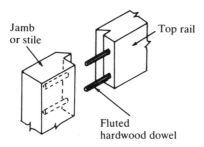

Fig. 9.6 Dowelled joint

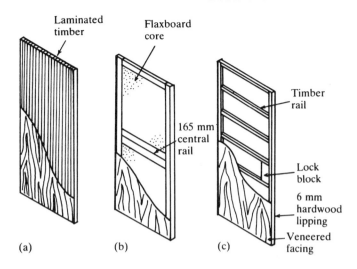

Fig. 9.7 Flush doors (a) Solid core (b) Semi-solid core (c) Hollow core

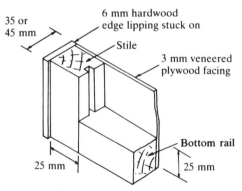

Fig. 9.8 Flush door construction joint

timber laminates will provide both excellent fire and sound insulation. Veneered plywood or hardboard facings are 3 mm thick and overall widths are 35 mm and 45 mm for internal and external use respectively; 6 mm hardwood edge lippings are also required to resist impact and abusive use. Figure 9.8 details a typical edge and corner joint between stile and bottom rail.

Fire check flush doors (see Fig. 9.9)

These doors are constructed to withstand the effects of a fire for a

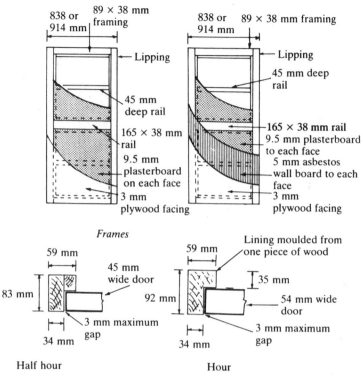

Fig. 9.9 Fire-check doors

designated period of ½ an hour or 1 hour. The choice depending on the building purpose and size. Normally, single- and two-storey dwellings are exempt, but three-storey construction will require at least half-hour fire resistance doors with an automatic closing device attached to each. Superficially they appear the same as ordinary flush doors. Internally the door is divided at mid height with a 165 mm central rail, and each half is filled with flaxboard or 9.5 mm plasterboard nailed to each side of a timber framework. With a 4 mm veneered plywood facing and an overall thickness of 45 mm, this construction satisfies the half-hour specification. For one hour, an additional layer of 5 mm asbestos partition board is bonded to each side of the flaxboard or plasterboard to give a minimum overall thickness of 54 mm. Two-hinge support and a planted stop lining are adequate for half-hour doors, whilst three hinges and a moulded lining must be used with one-hour doors.

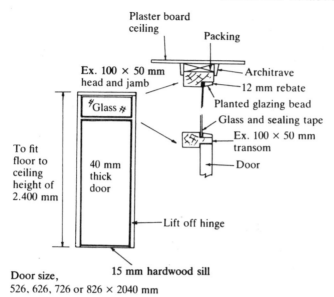

Fig. 9.10 Door set

Door sets

These are a combined frame/lining and door. They arrive on site with the door ready hung in a frame to save considerable site assembly time. A thin hardwood sill is necessary to complete the structural frame and an optional glazed or solid panel is available over the door to provide a storey-height frame. Figure 9.10 shows a typical factory produced unit with pin butts or lift-off hinges to allow easy door removal whilst the frame is fitted. Other door ironmongery is not normally supplied.

Windows

The functions of a window are to admit daylight, provide natural ventilation and to exclude rainwater. In some circumstances the view from a window provides an important function as relief and pleasant relaxation from the daily internal routine.

The materials for window production are principally timber and steel. Aluminium is a popular improvement on steel, having a rust- and maintenance-free life. Plastics (PVC) are now quite common, either as a coating to steel or timber, or in hollow extruded tubular

(A)   Window with minimum openable area = 1/20 floor area.
(B)   Mechanical ventilation of at least 3 air changes per hour.
(C)   Mechanical ventilation of at least 15 litres per second.
(D)   Mechanical ventilation of at least 60 litres per second (30 l/s if in cooker hood).
(E)   Trickle ventilation of at least 4000 mm^2.
(F)   Trickle ventilation of at least 8000 mm^2.

Note: The top part of a ventilation opening must be at least 1.75 m above floor level.

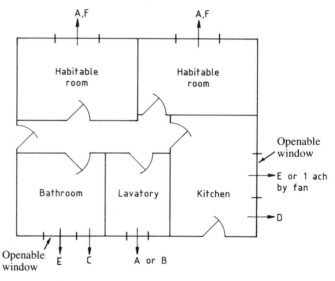

Fig. 9.11 Ventilation of dwellings

form as completed profiles. Corners are mitre cut and welded, and reinforced with steel brackets.

Windows cannot be installed as desired. The extent of glazing, ventilation and position is controlled by Parts F and L to the Building Regulations. Figure 9.11 is a summary.

Current energy conservation measures require windows to be double glazed with a *U* value not exceeding 3.0 W/m^2 K. Also, all window sashes and doors must be draught proofed. For dwellings, the maximum glazed area of windows and doors should not exceed 22.5% of the total floor area. Other residential buildings can have a maximum glazed area of 30% of exposed wall and sky lights may occupy up to 20% of the roof area.

The **Standard Assessment Procedure** (SAP) quantifies energy efficiency of new dwellings on a numerical scale 1 to 100. Approved Document L to the Building Regulations provides tables and charts to determine the value, considering the fabric U values, ventilation, efficiency of water heating and other energy transfer such as solar gains. The minimum acceptable is between 80 and 85, depending on the size of dwelling.

Alternatively, the **Elemental Method** may be used to show conformity with the Building Regulations. This will require a standard approach to construction, i.e.

(i)   roof insulation of 150−175 mm thickness,
(ii)   double glazing throughout,
(iii) brick and lightweight block exterior walls with a 75 mm insulated cavity,
(iv)   50 mm rigid insulation in ground floors.

Greater design flexibility can be achieved using the other alternative, the **Target U Value Method**. Deviations from standard construction are permitted provided the target U value from the following formulae exceeds the average U value.

$$\begin{array}{l} \text{Target } U \text{ value} \\ \text{where SAP} < 60 \end{array} = \left(\frac{\text{Floor area} \times 0.57}{\text{Area of exposed elements}} \right) + 0.36$$

$$\begin{array}{l} \text{Target } U \text{ value} \\ \text{where SAP} > 60 \end{array} = \left(\frac{\text{Floor area} \times 0.64}{\text{Area of exposed elements}} \right) + 0.40$$

See Fig. 9.12 for an example of application.

Element	Exposed area (m²)	U value	Heat loss (W/K)
Ground floor	40	0.35	14.0
Wall	39	0.40	15.6
Door/window	11	3.50	38.5
Roof/ceiling	40	0.20	8.0
	130		76.1

$$\text{Average } U \text{ value} = \frac{76.1}{130} = 0.585 \ W/m^2 \ K$$

$$\text{Target } U \text{ value} = \left(\frac{40 \times 0.64}{130} \right) + 0.4 = 0.596 \ W/m^2 \ K$$

Therefore the proposal is acceptable.

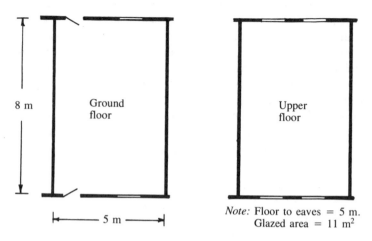

Fig. 9.12 Target 'U' value, eg. mid terrace house (SAP > 60)

Timber window types

The range of window patterns and styles now available to the purchaser is extensive, as joinery manufacturers compete to attract the house building and renovation market. Traditionally, timber windows were divided into classes according to the manner in which the sashes (glazed frame which fits into the window frame) were hung, viz:

(a) casement;
(b) double hung sliding;
(c) pivoted.

These classes are still valid as the basis for window design, but many modifications exist which are marketed under trade names such as sunshine, Georgian, Jacobean, high performance, etc. Overall frame dimensions are co-ordinated with standard bricks to preserve appearance and to avoid awkward cutting.

Traditional casement windows

Standard casement windows have an outer framework containing a head, sill and two jamb sections, and internal dividing bars known as a transom (horizontal) and mullion (vertical).

These contain deep rebates for sashes, which may be openable or fixed. Alternatively, the frame members may be glazed direct, although the appearance is less attractive. The sill is the most exposed component and should be made from hardwood. It is weathered with a 10° slope to the upper surface and grooved on the underside to provide a throating or drip and a mortar groove. The inside face contains a 10 × 10 mm recess for location of a window board. Figure 9.13 shows the outline of a typical casement window to BS 644, and Fig. 9.14 the construction details. The notation applied to these windows is adopted by joinery manufacturers for catalogue identification of their products. The example shown, 336V denotes:

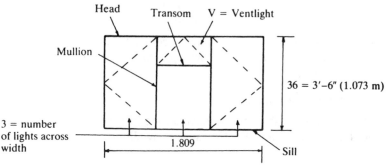

Fig. 9.13 BS casement window, ref. 336V

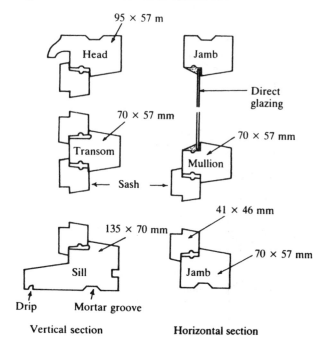

Fig. 9.14 Casement window sections

3 = No of lights in width

36 = The overall height (in feet and inches i.e. 3′ 6″)

V = Ventlight (small top hung light in centre)

Other standard notation:

N = Narrow light

P   = Plain (no transom or mullion)

T   = Through transom (uninterrupted from jamb to jamb)

S   = Sub-light (small deadlight above sill)

VS = Ventlight and sub-light

Note: Many manufactures have acquired additional notation symbols to suit their own products, e.g. PW = Picture window.

Sliding sash windows

Vertical sliding sash windows have a long history. They were very popular in Victorian and Edwardian dwellings, because the tall

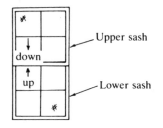

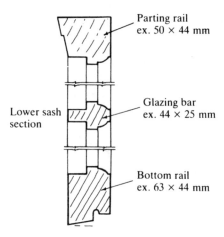

Fig. 9.15 Sliding sash details

relatively slender outline complemented the high ceilings associated with this period of construction. Figure 9.15 details sections through the sash, including the option of glazing bars. Figure 9.16 shows a horizontal section through the obtrusive jamb casing which contains the sash balance weights. These may be cast iron, lead or concrete and function with a sash cord passing through a small pulley wheel at the head of the boxing. This style of window has recently enjoyed a revival, but the operating mechanism shown in Fig. 9.17 has been simplified with a spiral spring balance to counter the sash weight. This eliminates the need

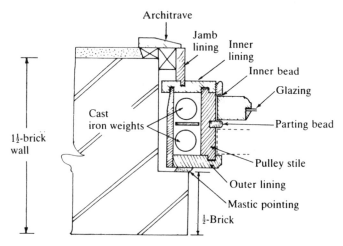

Fig. 9.16 Horizontal section through double hung sash window jamb

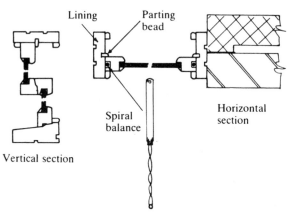

Fig. 9.17 Modern sliding sash window

for boxing at the jambs, providing a far more attractive window than the original.

Pivot windows

There is no British Standard for this window type, which would account for the variation between manufacturers. They are either horizontally pivoted with the hinges attached centrally to opposing jambs, or vertically pivoted between centre of head and sill. The former is the most common and is shown in outline and section in Fig. 9.18. These windows have the advantage of providing a pleasant appearance to the facade of a building and possession of easy access for cleaning and maintenance by sash rotation. The balanced effect also reduces the possibility of wind damage, although the effectiveness is often dependant on the frictional adhesion of the pivot hinge. Most manufacturers ensure that the hinge is recessed within the frame and sash, and covered with a weathering strip otherwise the security value with exposed ironmongery is questionable.

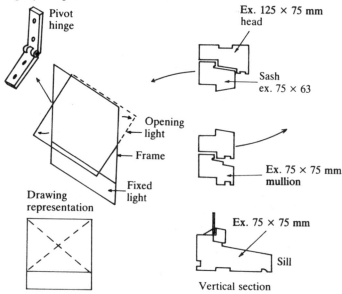

Fig. 9.18 Pivot window

Broken line notation

On architectural drawings and in catalogues it is customary to indicate the side of the window on which the sashes are hinged.

This is necessary on drawings to be presented for local authority approval and for ordering purposes. Casements are described as left or right hand depending on the side they are hinged as viewed from the outside. Figures 9.13 and 9.19 indicate inclined broken lines on the opening sashes which centralise against the hinging component. Pivoted sashes are shown with diagonal broken lines from each corner as shown in Fig. 9.18.

Steel windows

Modern steel windows can be manufactured to a dimensional space module of 100 mm, or are available in traditional imperial dimensions, and metric wall opening sizes that are multiples of 100 mm. The adopted width dimensions are:

500, 600, 800, 900, 1200, 1500 and 1800,

and height dimensions:

200, 500, 700, 900, 1100, 1300 and 1500 mm.

These are the wall opening sizes, the actual frame size is 6 mm less overall.

Standard sections are shown in Fig. 9.19. These are mitre (45°) cut at corners and welded, before receiving a final surface treatment of hot dip galvanising to prevent corrosion. Frames may be built into the wall with direct fixing from cramps or by attachment to a timber sub-frame of 75 × 50 mm or 75 × 75 mm section. These relieve the insignificance of the steel frame, but they do not fit into modular co-ordinated openings. Some examples are shown in Fig. 9.20. Notation for steel windows is similar to that used for timber frames, with the following abbreviations:

F   = fixed light
C   = side hung casement opening outwards
R   = Horizontally pivoted, reversible casement
V   = Top hung opening outward casement extending full width of frame.
T   = Top hung opening outward casement at side of C or F.
B   = Bottom hung casement opening inwards
S   = Fixed sub-light
D   = Casement door, opening outwards
LV = Deep top hung casement opening outwards.

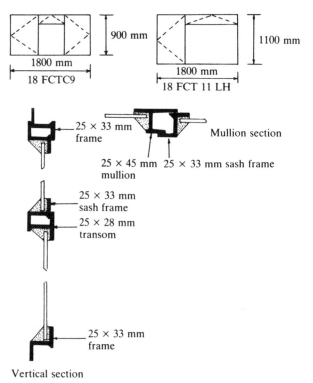

Fig. 9.19 Steel window notation and details

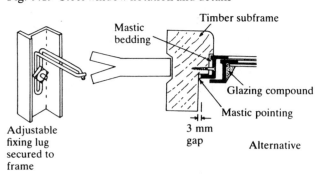

Fig. 9.20 Jamb fixing for steel windows

Also: (a) RH or LH may be used to denote a right-hand or left-hand hinged casement, as viewed from the outside;

(b) The prefix number refers to the quantity of 100 mm horizontal modules in the frame width, and the suffix number, the quantity of 100 mm **vertical modules.**

Examples: 18FCT 11 LH

>1800 mm wide × 1100 mm high
>One fixed light
>One side hung left hand casement
>One top hung casement

>18FCT C9

>1800 mm wide × 900 mm high
>One fixed light
>Two side-hung casements
>One top-hung casement

Glass and glazing

Glass is made by fusing sand (70%), soda (13%), lime (13%) and other minor ingredients such as metal oxides and clay at a temperature of 1500 to 1550 °C. The molten glass is drawn vertically for about 10 m and then along horizontal rollers where it is cut into sheets. Patterned glass is made in the same manner, but reduced in size by a series of opposing steel rollers. The final rollers are surface profiled to produce a patterned effect on one side of the glass. Both drawing and rolling are used for making clear glass, but the result is never perfect as the strain of drawing and the mark from rollers produce slight surface flaws and deformities. Perfectly finished glass for mirrors, cabinets and other high glass joinery is produced by floating the molten glass over a bath of molten tin until all the imperfections even out. For general glazing purposes glass is produced in 3, 3.5 and 4 mm thicknesses.

Transparent glass. Transparent or clear sheet glass has a light transmittance efficiency of about 85%. When used as double glazing this figure reduces to about 70%. Three grades of glass are specified for normal use:

1 OQ, ordinary glazing quality for use in most construction work for doors and windows.
2 SGQ, special glazing quality for use where the client requires a superior finish to glazed components.
3 SSQ, special selected quality used for especially finely finished situations. This may be obtained by grinding and polishing rolled glass or by float process.

Glazing

Various examples are shown in Fig. 9.21. These include the use of:

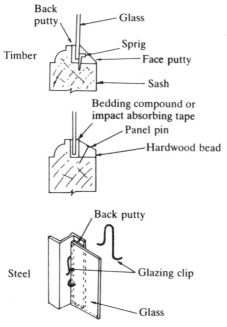

Fig. 9.21 Sash glazing: timber and steel

(a)   linseed oil putty;
(b)   metal casement putty;
(c)   hardwood beading.

In all situations the glass should be cut to allow a 2 mm clearance between pane and sash/frame.

Putty

Linseed oil putty is for use with a primed timber backing. It is a mixture of linseed oil and whiting and dries by oxidisation and absorption into the timber background. Metal casement putty is made from refined vegetable drying oils and chalk, which dries rapidly by evaporation of the oil. It must be specified for metal frames, as in this situation the non-absorbant background would prevent linseed oil putty from drying. For both timber and metal components, a 2 mm layer of back putty is applied to the sash before placing the glass. The glass is retained with sprigs (small nails with a wide head) in timber and spring clips in metal at 300 to 450 mm spacing, before applying the face putty. Face putty is positioned with the thumb and struck off at an angle with a purpose-made putty knife.

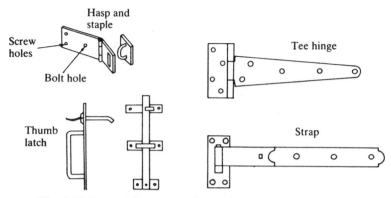

Fig. 9.22 Ironmongery – matchboard doors

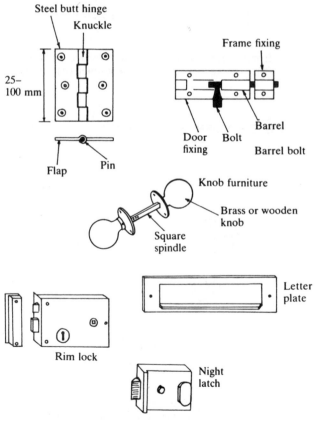

Fig. 9.23 Ironmongery – panelled doors

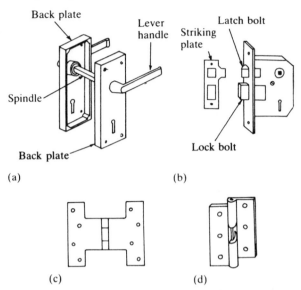

(a)　　　　　　　　　　　　　　(b)

(c)　　　　　　　　　　(d)

Fig. 9.24 Ironmongery – Flush doors (a) Mortice lock furniture
(b) Mortice lock (c) Parliament hinge (d) Rising butt

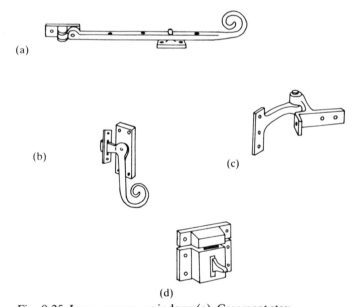

(a)

(b)　　　　　　　　　　(c)

(d)

Fig. 9.25 Ironmongery – windows (a) Casement stay
(b) Casement fastener (c) Easy clean hinge (d) Fanlight
fastener

Beads

Glazing beads are generally hardwood and attached to the inside of a sash with panel pins or cups and countersunk screws. The glass is placed in a glazing compound or edged in an impact absorbing tape as shown in Figs 9.4 and 9.21.

Ironmongery

The range of ironmongery for doors and windows is extensive. Price variation is also considerable, depending on the material selected, and functional mechanism (particularly with locks) and the finish. Examples to suit the doors and windows discussed in this chapter are shown in Figs 9.22 to 9.25.

Draught proofing

Conservation of fuel and improved interior comfort are achieved by sealing breaks where construction is discontinuous. Joinery manufacturers now provide brush or plastic compressible seals as standard accessories to opening components in windows and doors. Rooflights and loft hatches are also suitably sealed. Penetration and interruption of the structure by services and window or door frame junctions, plus gaps between dry lining and masonry walls will require attention from the builder during construction. Contemporary mastic and silicon products are most appropriate in these situations.

10

Stairs

A stairway is initially designed to provide an effective means of access between different floor levels. A secondary function, of considerable importance, is to provide a practical escape route in the event of fire.

The most popular materials for stair construction are timber and reinforced concrete. Timber is used for virtually all domestic stairs, because of its economic availability, adaptability and compatibility with the associated construction. In flats, offices, factories, etc., concrete is preferred because of its durable qualities and greater resistance to fire.

Domestic timber stairs

A stairway in a house must be designed for safe use by occupants of various ages and physical capacity. Therefore, unless it is purpose-made, the stairway used in most dwellings is a form of construction which compromises between the requirements of the most fit and the least agile, and the tallest and the shortest. These requirements are legalised in Part K to the Building Regulations. This section imposes strict control over dimensional tolerances for components, to ensure that in normal use an accident is impossible. Should misuse cause an accident, built-in safety features, e.g. handrail, maximum pitch, provide for correction. Figure 10.1 summarises the controls affecting stair construction, which include the following:

- Equal rise for every step or landing
- Equal going for every parallel tread
- The maximum pitch angle to the horizontal is 42°
- Going of a tread, at least 220 mm
- Rise of a tread, at least 155 mm and no greater than 220 mm

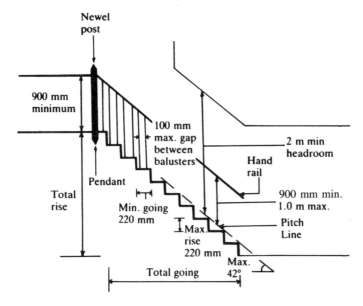

Fig. 10.1 Legislation governing domestic stairways

- Headroom measured vertically above the pitch line is at least 2 m
- The sum of twice the rise plus the going is equal to, or between 550 mm and 700 mm
- Stair length not greater than 16 rises
- Handrail provided at a height between 900 mm and 1 m above the pitch line
- Balusters spaced to prevent a 100 mm diameter sphere passing through

Additional relevant Building Regulations apply to open tread stairs and spiral or tapered tread stairs.

Open tread stairs

The term open tread is confusing, because if the tread were open there would be nothing to stand upon! It of course refers to the open space below each tread, which must not exceed 100 mm as shown in Fig. 10.2. Also shown is the minimum dimension of

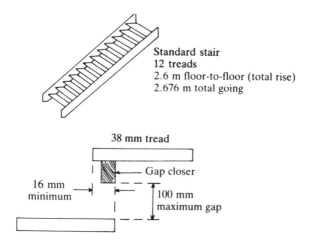

Standard stair
12 treads
2.6 m floor-to-floor (total rise)
2.676 m total going

Fig. 10.2 Open-tread stair

16 mm. This represents the nosing overlap of every tread with the tread below.

Tapered tread stairs

Tapered treads are a useful method for turning a straight flight through 90° with a simultaneous gain in height. The safety of the stairs is questionable as the tapered treads break the rhythm of ascent or descent, particularly at the coincidence with the newel post. Here the going is dangerously small, and must be at least 50 mm to satisfy the Building Regulations. Spiral stairs are an extended version of tapered treads. Many manufacturers take advantage of the aesthetic potential of these stairs and produce some very elaborate and attractive designs. However, in the interests of safety and accessibility for furniture, a straight flight must be preferable. The following legislation must be measured at a point midway along the tread length as shown in Fig. 10.3:

Going at least 220 mm
Twice the rise + going, at least 550 mm and not greater than 700 mm
Pitch angle at least 42°

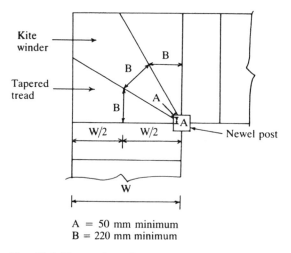

A = 50 mm minimum
B = 220 mm minimum

Fig. 10.3 Tapered treads

Timber stair construction

The most convenient stair arrangement is a straight flight, with one side fixed directly onto a wall and the other side presenting the string, balusters and a handrail. These stairs are purpose-made by joinery manufacturers to suit 2.6 m (12 treads) or 3.0 m (14 treads) floor-to-floor dimensions. This and other stair layout plans are shown in Fig. 10.4. Provision of landings is necessary in old persons' homes, where the number of rises exceeds sixteen and where the stair is to change direction. This latter provision is a more compact arrangement which will often benefit room layout, but the additional components and more difficult fitting will add to the cost.

Very few stairs are now constructed *in situ* because of the time delay and associated labour costs. Figure 10.5 shows a view to the underside of a purpose made or ready made stair with treads and risers housed into the strings, located firmly with wedges. Treads and strings are usually 32 mm thick Parana pine or dense fibreboard, and risers 16 mm plywood. Treads and risers are tongued and grooved together, and given additional stiffness with at least three triangular glue blocks. The upper end of the wall string is cut and notched over the trimming member. The outer string has both upper and lower ends housed in newel posts with oblique haunched double tenon joints, as detailed in Fig. 10.6. Newel posts also support the handrail and in turn receive direct bearing support from the lower floor.

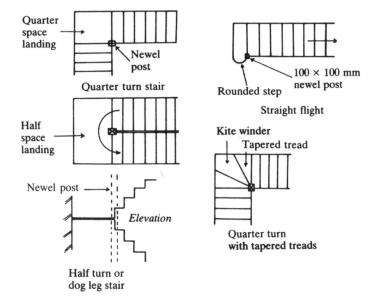

Quarter
space
landing

Newel
post

Quarter turn stair

Half
space
landing

Newel post

Elevation

Half turn or
dog leg stair

Rounded step

100×100 mm
newel post

Straight flight

Kite winder
Tapered tread

Quarter turn
with tapered treads

Fig. 10.4 Stair profiles

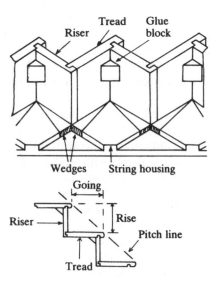

Riser   Tread   Glue
                  block

Wedges   String housing

Going

Riser   Rise

Pitch line

Tread

Fig. 10.5 Housing of treads and risers

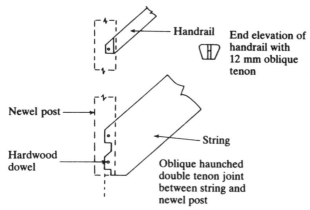

Fig. 10.6 Newel post construction joints

Alternatively, the upper newel may have a 20 mm notch and coach bolt connection to the upper trimming timber. In this situation the newel post is terminated below ceiling level with a decorative pendant moulding to match the top.

Balustrade (wall, screen or railing)

A handrail is positioned between 900 mm and 1 m above the pitch line to run parallel with the string. Around landings and open edges of floors the handrail height is at least 900 mm. The space between handrail and string may be completely closed by timber studwork and a boarded cladding, but this is not as attractive as using plain or preferably turned spindles, otherwise known as baluster rails. These house into the handrail and string or a string capping where the additional feature is required. Figure 10.7 shows possible treatment here.

Apron lining

To improve the appearance around the stair opening in an upper floor, a rebated nosing planed to match the treads is fitted to the floorboards and trimming timber as shown in Fig. 10.8. A piece of planed softwood known as an apron lining is tongued along the upper edge to fit the underside of the rebated nosing. Strips of packing timber are usually required between lining and trimming to locate the lining accurately. The junction between apron lining and plasterboard ceiling is covered with a moulding, similar to that used as an architrave around door openings.

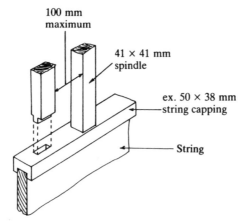

Fig. 10.7 String capping

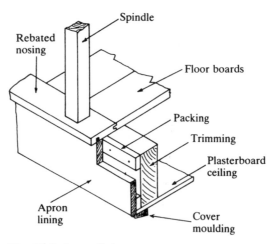

Fig. 10.8 Apron lining

Rounded bottom step

This is sometimes referred to as a built-up curtailed riser because of the manner in which it is produced. This optional feature is time consuming to make and therefore an expensive attraction to the bottom of a stair. It is manufactured by cutting a veneer from the riser board and wrapping this around a built up glued and screwed block as detailed in Fig. 10.9. Folding wedges are also required to ensure a tight location of veneer to blocks.

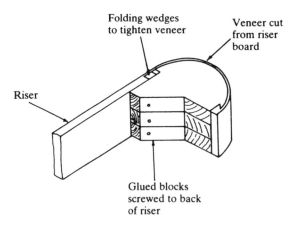

Folding wedges
to tighten veneer

Veneer cut
from riser
board

Riser

Glued blocks
screwed to back
of riser

Fig. 10.9 Rounded bottom step

Reinforced concrete stairs

A simple reinforced concrete stair has a similar structural behaviour to a simply supported floor slab. For design purposes the effective span is measured between the centres of opposing supports and the thickness taken as the waist. Figure 10.10 illustrates these factors and a more efficient design using downstand beams to reduce the effective span and bending moment. In both examples starter bars are shown extending out of the floor slab to provide reinforcement continuity. These are tied with wire to the main reinforcement. The pitch angle of 38° maximum is also shown, this applies to all public buildings. There are many other dimensional variations with domestic private stairways which are not detailed, due to the large variety of building types and related purpose groups which each have specific regulations. The reader should refer to Part K of the Building Regulations for determination of the requirements for any particular public building. Concrete should be very strong of compressive strength 25–30 N/mm^2 conforming to a mix ratio of 1:1½:3 with coarse aggregate graded from 10 mm down to 3 mm. Watercement ratio is low (0.5 max) otherwise concrete will flow over the formwork riser boards.

Formwork

Stair formwork differs considerably from the remainder of the floor formwork. The underside contains a 25 mm inclined plywood sheet supported on joists and struts as shown in

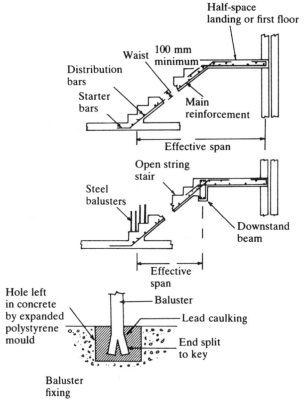

Fig. 10.10 Simple reinforced concrete stairs

Fig. 10.11. The steps are formed by securing boards to the adjacent walls and suspending cleats and riser boards at the required spacing. Where walls do not occur, an open string may be formed by using edge formwork cut to the stair profile as shown in Fig. 10.12. Reinforcement spacers should be used to provide at least 20 mm concrete cover.

Precast reinforced concrete stairs

As can be seen from the previous section, stairs cast *in situ* are considerably more difficult to create than columns, beams and floors. The irregular shape and inclined soffit create difficulties which consume considerable formwork production time. Precast stairs are compatible with precast floor systems, particularly as lifting equipment is on site and as the need for formwork would

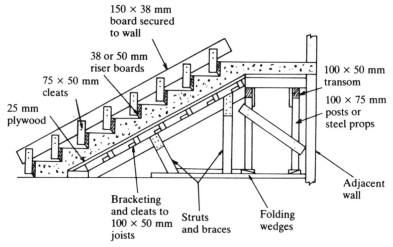

Fig. 10.11 Formwork to concrete stairway

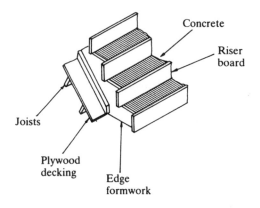

Fig. 10.12 Formwork to open string stair

break the construction routine by requiring carpenters at intermittent stages. Where precast stairs are used with *in situ* concrete floors a recess is left in both top and bottom floor slabs to accommodate a step left in the stair. Figure 10.13 illustrates a typical half flight stair for use with a landing, including reinforcement scheduling details, and Fig. 10.14 shows how the floor is rebated and dowelled for stair positioning.

Precast concrete steps

These may be used as an entrance feature with direct bearing on

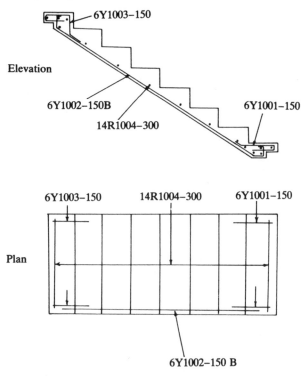

6Y1003–150

Elevation

6Y1002–150B

14R1004–300

6Y1001–150

6Y1003–150   14R1004–300   6Y1001–150

Plan

6Y1002–150 B

Fig.10.13 Precast concrete half flight of stairs: elevation and plan

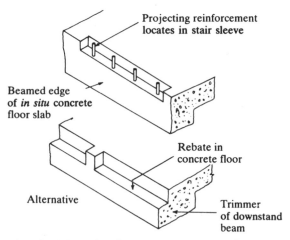

Projecting reinforcement
locates in stair sleeve

Beamed edge
of *in situ* concrete
floor slab

Rebate in
concrete floor

Alternative

Trimmer
of downstand
beam

Fig. 10.14 Location for precast concrete stair

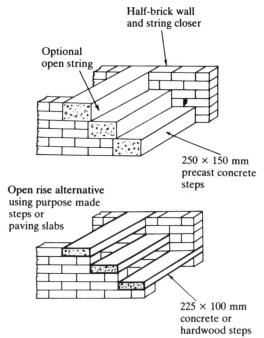

Half-brick wall
and string closer

Optional
open string

250 × 150 mm
precast concrete
steps

Open rise alternative
using purpose made
steps or
paving slabs

225 × 100 mm
concrete or
hardwood steps

Fig. 10.15 Precast concrete steps

the ground, as shown in Fig. 10.15. Here, support is at either end
and reinforcement minimal. An alternative method for longer
flights is to build one end of the steps into the wall and to support
the other ends on a steel or prestressed concrete beam as shown in
Fig. 10.16.

Spandrel cantilever steps

These steps are built into a wall at one end only and receive no
outer string support. As only one end is supported, a minimum of
225 mm or one brick thick wall hold is necessary. Temporary
support during construction will be required at the free end as
shown in Fig. 10.17. Figure 10.18 shows sectional details and
dimensions of these steps which contain at least two 16 mm
diameter steel reinforcing rods close to the upper surface to resist
the bending stresses imposed by the cantilevered situation. An
alternative application uses precast tapered treads centred on a
steel tube to create an open riser spiral stair. This is shown in
Fig. 10.19 with two 16 mm diameter steel balusters on every tread
to support a tubular steel handrail.

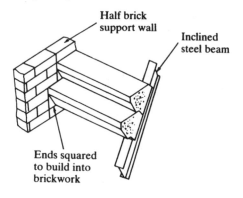

**Half brick
support wall**

**Inclined
steel beam**

**Ends squared
to build into
brickwork**

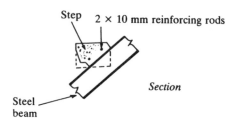

Step   2 × 10 mm reinforcing rods

Section

Steel
beam

Fig. 10.16 Long flight precast concrete steps

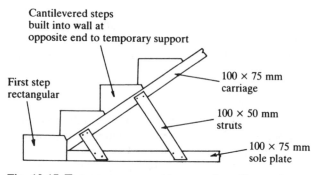

Cantilevered steps
built into wall at
opposite end to temporary support

First step
rectangular

100 × 75 mm
carriage

100 × 50 mm
struts

100 × 75 mm
sole plate

Fig. 10.17 Temporary support to precast cantilevered steps

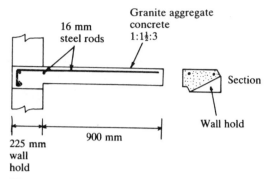

Fig. 10.18 Cantilever step detail

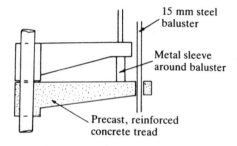

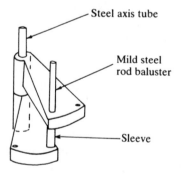

Fig. 10.19 Open riser spiral stair

11

Internal and external finishings

Internal finishes

Floors

The range of finishes available is extensive, as is the quality and cost of installation. Selection of a suitable material should depend on the building or room function with regard to durability, appearance, effectiveness, comfort and anticipated maintenance. Carpeting is a good example of cost and quality variation. Traditional woven carpets such as Axminsters and Wiltons have the pile woven with the backing and into the backing respectively. A cheaper, popular alternative is tufted carpet, where the pile is stitched into a jute backing and stuck with a latex adhesive. Pile materials vary between the very costly wool fibres which have excellent wearing qualities and the choice of several man-made fibres such as acrylic, nylon, polypropylene and rayon. Many manufacturers have experimented with mixtures of synthetics and natural fibres to achieve moderately priced, hard wearing carpets.

Carpeting has rarely featured as part of the building contractor's work. In recent years, as competition between some housing developers has increased, the fully-fitted house has included carpets and kitchen machinery to improve the selling potential. It is improbable that the builder would employ carpet fitters as direct labour, and this work is usually subcontracted to the most competitive local installer.

The following includes a few examples of floor finishes more appropriate to builders' work.

Softwood boards

These are the traditional finish to suspended timber ground and upper floors. In older housing, subject to renovation and improvement, sanded and varnished boards offer an attractive and

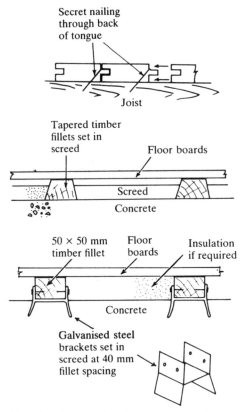

Fig. 11.1 Tongued and grooved floor boarding

hardwearing surface where the expense of superficial covering is unwanted or undesirable. As a floor structural component, boards have been superseded by chipboard or plywood mainly for reasons of cost, although it is preferable to use two-dimensional load distributing sheet materials for structural reasons also. Where softwood boards are specified as part of the floor structure and as the finish, they should be tongued and grooved to eliminate draughts and to facilitate secret nailing as shown in Fig. 11.1. Also shown is the application of boards to concrete sub-floors, using brackets or timber fillets attached to the concrete.

Hardwood block and strip

Contemporary parquet block floor finishes are produced in 600 mm squares comprising sets of 100 mm long × 6 mm thick hardwood strips arranged in basketweave patterns. The strips are attached to a flexible backing such as jute or foil and stuck direct to a prepared screed or boarded floor. Some manufacturers provide the strips already stuck to a rigid plywood backing which is useful if the sub-floor has slight surface irregularities. Traditional hardwood (usually oak) block floors are an expensive but very attractive hardwearing floor finish, used occasionally for entrance halls to prestige buildings. They are often found in school halls/gymnasiums because of their excellent wearing qualities. Figure 11.2 shows the format of these blocks which tongue and groove or dowel together to prevent individual movement. They are secured to a screeded sub-floor with hot or cold bituminous solution. Final treatment is by sanding and several layers of varnish.

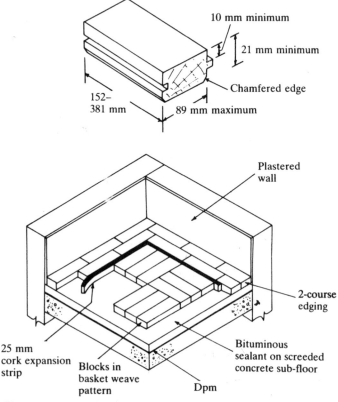

Fig. 11.2 Hardwood block flooring

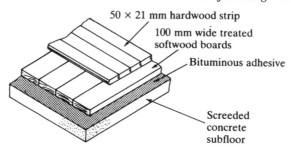

50 × 21 mm hardwood strip

100 mm wide treated softwood boards

Bituminous adhesive

Screeded concrete subfloor

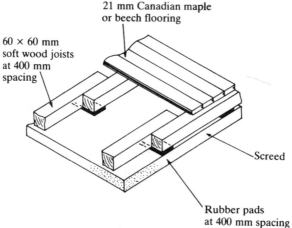

21 mm Canadian maple or beech flooring

60 × 60 mm soft wood joists at 400 mm spacing

Screed

Rubber pads at 400 mm spacing

Fig. 11.3 Gymnasium/dancehall floors

Hardwood strips for floor finishes to gymnasiums, dance floors, etc., are available in nominal widths up to 100 mm. Lengths are random, and sections are joined together with tongued and grooved edges and ends, with secret nail location to an underlay of softwood boarding on a screeded or power float finish concrete structural floor. Figure 11.3 shows this arrangement and a variation for sports halls using resilient rubber pads under supporting joists to reduce the effect of personal impact, whilst still retaining sufficient ball response.

Plastic tiles

These originated as a thermoplastic or asphalt tile secured to a screeded floor with a hot bituminous solution. Although many installations are now over forty years old, the hardwearing qualities of these tiles have stood the test of time. The main weakness is brittleness and hard finish which has caused fracture and displacement of sections of tile, particularly where there has been some movement in the floor. Flexible PVC tiles have superseded these rigid tiles and provide for more accommodating properties for simple cutting, placing and adhesion with synthetic glues. Colours and pattern potential is considerable, and maintenance is much easier once the surface is sealed and polished with wax. BS 3261 specifies thicknesses of 1.5, 2, 2.5 and 3 mm. Overall sizes are 225, 250 or 300 mm square. A cheaper plastic tile containing fillers of asbestos fibre and limestone is manufactured to BS 3260. These have most of the properties of PVC tiles but are more brittle, less resistant to marking and more difficult to clean and maintain.

Clay or quarry tiles

Clay tiles or 'quarries' are manufactured by cutting extruded refined clay into BS 1286 preferred dimensions of 200 × 100 mm or 100 × 100 mm × 9.5 mm thick. Thicker 'quarries' are made by compressing clay into moulds of up to 19 mm depth. They are dense, impervious, very durable and easy to clean, properties attractive for use in canteens, food processing factories, lavatories, etc. They are often used as a practical but attractive feature in domestic porches and entrance lobbies. Colour depends on the clay source, dull red and yellow being the most common. Fixing is shown in Fig. 11.4 with a 20 mm thick cement and sand bedding of 1:4 ratio and a separating layer of polythene under the bedding to permit movement, which is absorbed by the cork expansion joint.

Wall and ceiling finishes

Plaster

Building plasters are produced from gypsum rock. This is crushed, heated and ground to a fine powder before bagging and delivery to site. On site, water is reintroduced to the powder to create a plaster solution for application to walls and ceilings before a chemical reaction and evaporation occur, which return the solution to the original gypsum rock. Plaster composition, quality and finish vary to suit different applications. Table 11.1 gives a list of plaster classifications in accordance with BS 1191, which omits

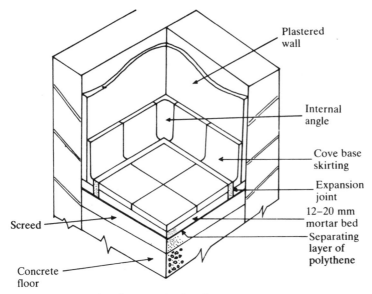

Plastered
wall

Internal
angle

Cove base
skirting

Expansion
joint

12–20 mm
mortar bed

Separating
layer of
polythene

Screed

Concrete
floor

Fig. 11.4 Clay/quarry tile flooring

non-standard plasters for specialist applications such as X-ray protection and acoustic insulation.

Types of plaster and application

Class A: Plaster of Paris. A quick-setting plaster, only mixed in small quantities for minor repairs and crack filling.

Class B: Retarded hemihydrate. A coarser gypsum than plaster of paris with a retarder added to slow the setting time. This permits large quantities to be mixed which remain workable for at least 20 minutes. Type (a) undercoats are generally associated with rendered or sandy backgrounds and type (b) is a finishing plaster which may be gauged with 25% lime. Plasterboard-finishing plaster is specially formulated for the paper-faced background and is applied in a thin skimming coat to mask background imperfections and joints.

Class C: Anhydrous. This is produced by total dehydration of gypsum before bagging and delivery to site. The bonding quality is poor, except to cement and sand rendered undercoat and gypsum plaster undercoats. The setting and hardening process is slow, which coupled with the poor background qualities, limit this plaster to finishing work only.

Table 11.1

Gypsum-based plasters	British Standard	Trade name
Class A Plaster of Paris hemihydrate	BS 1191: Part 1	CB Stucco
Class B Retarded hemihydrate undercoats:	BS 1191: Part 1	
browning	a1	Thistle browning Thistle slow setting browning Thistled fibred
metal lathing	a2	Thistle metal lathing
Finishes:		
finish	b1	Thistle finish
board finish	b2	Thistle board finish
Class C Anhydrous finish only	BS 1191: Part 1	Sirapite
Class D Keenes	BS 1191: Part 1	Standard Keenes Fine, polar white cement
Pre-mixed lightweight undercoats:	BS 1191: Part 2	
browning	a1	Carlite browning Carlite browning (HSB)
metal lathing	a2	Carlite metal lathing
bonding	a3	Carlite bonding Carlite welterweight bonding
Finishes:		
finish	b1	Carlite finish Limelite finishing

Class D: Keene's. This is also a finishing grade plaster with greater purity and hardness, suitable where the surface may be subject to abrasive treatment.

Background

The plastering process will depend considerably on the uniformity and level of wall and the density, relative to the suction effect. Irregularities of the above will create the need for three-coat work, rendering (10–12 mm), undercoat (6–8 mm) and finish (2–3 mm). The other extreme, a smooth, dense surface such as plasterboard, will require one finishing coat only. However, finishing grade plasters will not adhere very successfully unless the surface is pretreated with a PVA bonding agent or a cementitious slurry brushed on. Plasterboard is excepted pretreatment, because a successful chemical bond is achieved between board and the specially formulated Class B (b2) finishing grade plaster. Steelwork and painted or glazed surfaces are other examples of difficult surfaces. These again, may be pretreated with a bonding agent or alternatively a surface preparation of galvanished wire mesh or expanded metal lathing bound to the steel and nailed to walled backgrounds. Undercoat in these situations is a Class B lathing grade plaster usually mixed with sand in the ratio of $1:1\frac{1}{2}$ and applied with an 8 to 10 mm thickness, followed by 2–3 mm of Class B or C finishing plaster. Softer backgrounds such as cork, expanded polystyrene or fibreboard are also prepared with mesh or expanded metal lathing stapled at 100 to 150 mm spacing. Undercoat is a lightweight bonding plaster (BS 1191:Pt.2:a3) with a compatible finish, (BS1191:Pt.2:b1).

Corner treatment

Internal corners should be reinforced with jute scrim to prevent shrinkage cracks, particularly at the intersection of plasterboard ceiling and wall. An alternative which relieves the sharpness of this corner is a gypsum cove, nailed or glued in position as shown in Fig. 11.5. Also shown are expanded metal angle beads secured to external corners with dabs of plaster before finishing just below the nosing level.

Plaster application

Plaster is mixed with water until a workable consistency is obtained. Plywood boards of about 1 m square are often used as a mixing surface. When raised from the ground on trestles, this is known as a spotboard. An alternative for mixing purposes is an old galvanised steel bath, but these are now rare, and highly valued by plasterers. Various tools are shown in Fig. 11.6. The trowel or float is used to transfer small quantities of plaster to the hawk and then to the background. All final coats are steel trowelled to a smooth finish, whilst undercoats are scored with the edge of a steel

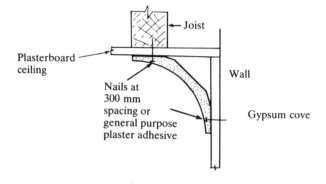

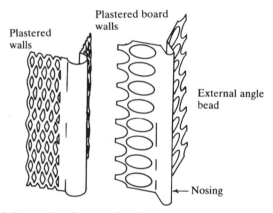

Fig. 11.5 Internal and external angles

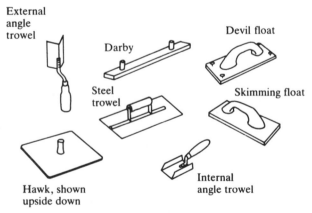

Fig. 11.6 Plastering tools

trowel or with projecting nails from an old wooden skimming float, known as a devil float, to provide a key for subsequent layers. The darby is simply a timber straightedge for levelling plaster on irregular backgrounds.

Plasterboard and drylining

Plasterboard lining to ceiling and timber stud walling has been standard practice since the demise of timber strips or laths and plaster in pre-war construction. In recent years it has been accepted as a quick and simple method for finishing brick or block masonry walls, instead of traditional wet applications of render and plaster which cause drying-out delays. Gypsum plasterboard is produced to BS 1230 in several different forms:

1 *Wallboard*. This is produced with two grey papered surfaces or one grey and one ivory coloured paper surface. The former is designed to receive a skim coating of Class B board finish plaster on either face and the latter, if fixed with the ivory surface exposed can be painted or papered directly, without pre-plastering.

2 *Lath*. Laths are produced in a narrow width and short length for easy access and fixing in difficult situations, such as ceilings. One surface only is prepared for plastering, the other is clearly marked, 'this side to joists', to avoid confusion. Edges of laths are distinctly rounded.

3 *Baseboard*. This is the same length as laths but of greater width. Edges are square and either side may be exposed for plastering.

4 *Plank*. Surface finish is the same as wallboard, with the option of two grey faces or one grey and one ivory for decorating direct. Planks are produced in narrower widths and greater thickness than wallboard for fire cladding to structural elements and lining to doors and partitions, where sound insulation or fire protection are required.

5 *Foil backed*. Foil backing is an optional facility, applied to all plasterboards to help reduce heat loss and to provide a vapour barrier for timber-framed inner leaf construction.

Table 11.2 compares the dimensions of the plasterboards discussed, and Fig. 11.7 shows edge treatment and non-standard thermal board of 22 to 65 mm overall thickness. Many other non-BS boards are manufactured for specific applications, these include aerated gypsum (insulation), glass fibre reinforced (fire) and polythene backed (vapour barrier).

Table 11.2   BS Boards

Type	Thickness (mm)	Width (mm)	Length (mm)
Wallboard	9.5, 12.5	600, 900, 1200	1800, 2350, 2400, 2700, 3000
Lath	9.5, 12.5	406	1200
Baseboard	9.5	914	1200
Plank	19	600	2350, 2400, 2700, 3000

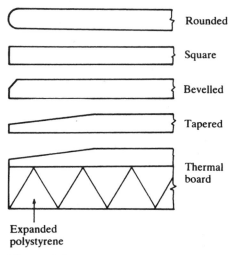

Rounded

Square

Bevelled

Tapered

Thermal board

Expanded polystyrene

Fig. 11.7 Plasterboard edge finishes

Plasterboard lining. Walls may be lined with plasterboard by (a) nailing the board to firmly fixed timber battens or (b) bonding the board onto the wall with plaster dabs.

(a) *Timber battens.* This technique is illustrated in Fig. 11.8, where battens are shown at the tapered edges of the board and in the centre. Horizontal joints should be avoided and battens are only required at the top and bottom to provide rigidity, particularly behind the skirting board. Nailing is with galvanised steel taper-head nails at 150 mm spacing as detailed with batten centres in Table 11.3.

(b) *Plasterboard bonding.* This is shown in Fig. 11.9 with bitumen-impregnated fibreboard pads stuck to a brick or block background with plasterwork general purpose adhesive. Top pads

Table 11.3 (All dimensions in millimetres)

Board thickness	Board with	Battern centres	Nails
9.5	900	450	2×30
	1200	400	
12.5	600	600	2×40
	900	450	
19	1200	600	2.6×60
	600	600	

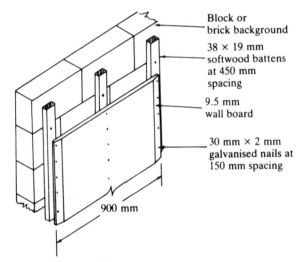

Block or
brick background

38×19 mm
softwood battens
at 450 mm
spacing

9.5 mm
wall board

30 mm $\times$ 2 mm
galvanised nails at
150 mm spacing

900 mm

Fig. 11.8 Dry lining to battens

should be about 230 mm from the ceiling and the lower pads 200 mm from the floor. After the pads are aligned and firmly secured, dabs of bonding or board finish plaster are trowelled on between them. The board is pre-cut about 13 mm short of the floor to ceiling height and placed firmly over the dabs. It is temporarily held with duplex or double headed nails in the pads until the plaster dabs have set. Nails are removed and reused.

Joint treatment. Figure 11.10 shows the standard joint between two tapered edge boards, using galvanised nails for fixing, a gypsum filler and reinforcing tape. The 3 to 5 mm gap and trough between the boards is filled and while still moist the reinforcing tape is pressed into the filler to bind both boards. A furth-

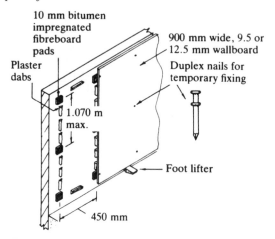

10 mm bitumen
impregnated
fibreboard
pads

Plaster
dabs

900 mm wide, 9.5 or
12.5 mm wallboard

Duplex nails for
temporary fixing

1.070 m
max.

Foot lifter

450 mm

Fig. 11.9 Plasterboard dry lining

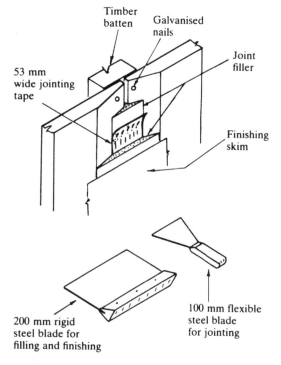

Timber
batten

Galvanised
nails

Joint
filler

53 mm
wide jointing
tape

Finishing
skim

200 mm rigid
steel blade for
filling and finishing

100 mm flexible
steel blade
for jointing

Fig. 11.10 Plasterboard joint treatment

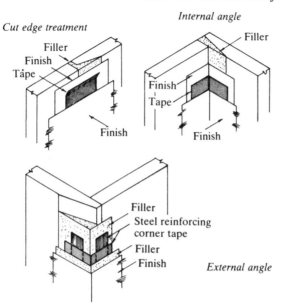

Fig. 11.11 Square edge and corner treatment

er layer of filler is applied to level the joint which is finished with a very dilute layer of finishing compound to eliminate surface irregularities. Internal corners and cut edges are treated similarly with reinforcing tape bonded with finishing compound as shown in Fig. 11.11. Also shown is an external corner. This requires pre-filling and steel reinforcing corner tape to accurately align and strengthen the corner before final filling and finishing.

Glazed ceramic wall tiles

Ceramic wall tiles are a similar product to clay floor tiles, using kiln-fired earthenware with the addition of glazed treatment to one surface and exposed edges. Sizes for imperial or non-modular tiles are 152 × 152 × 5, 5.5, 6 or 8 mm thick and 108 × 108 × 4 or 6.5 mm thick. Modular sizes, measured centre-to-centre of joint to incorporate spacer lugs as shown in Fig. 11.12 are 100 × 100 × 5 mm or 200 × 100 × 6.5 mm. Adhesion to the wall may be obtained with a 6–10 mm thick bedding of cement and sand mortar no stronger than a 1:3 ratio. This is often used for irregular backgrounds, otherwise a thin layer of mortar of about 3 mm thickness will be sufficient. Mastic ceramic tile adhesives are usually preferred for modern tiling as these will absorb modest

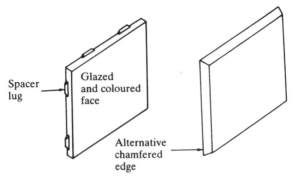

Fig. 11.12 Glazed ceramic tiles

background movement without displacing the tiles. Joints are grouted with a white cement solution, sponged or worked into the joint with the fingers. After about an hour the joints may be wiped clear of surplus grout with a damp cloth. Epoxide based grouts are used in hospitals and food preparation areas as these are impervious. They should not be applied with the fingers, but worked into the joint with a small knife or similar instrument.

Wallpaper

Lining paper. This is a cheap, plain white pulp paper used for preparing uneven and cracked surfaces before final papering. It reinforces old lime plastered walls and ceilings, and improves the uniformity of surface suction. A stronger brown paper is available for very fragile surfaces, but if the background condition has deteriorated this far, replastering or plasterboard lining must be seriously considered. Lining paper is available in rolls 11 m long × 560 or 760 mm wide. It should be applied at right angles to the finish papering to avoid the coincidence of joints; this is known as cross-lining.

Decorative paper. Material quality and composition vary considerably, which has a similarly variable influence on price. Some of the more popular papers are listed below. These are normally available in standard lengths of 10 m × 530 mm width, but effective length will depend considerably on the pattern repeat. This is where the pattern coincides on adjacent sheets, and it may have up to a 2 m deviation, leading to considerable wastage.

(a) *Flock.* A raised pile in patterned form, produced by blowing soft cotton or nylon fibres over an adhesive print.

(b) *Metallic.* An effect achieved by adhering metal powders to a paper background as flock papers, or by coating the paper with metal foil.

(c) *Ingrained.* Wood chippings, sawdust or fibres sandwiched between two fine layers of paper or adhered to the surface for emulsion painting.

(d) *Textured and relief papers.* Embossed rollers transfer their pattern to produce an imitation embossing on the paper. More expensive papers of better quality are lincrusta and anaglypta. Lincrusta has a low profile. It has a composition of whiting, wood flour, lithopone, wax, resin and linseed oil which when hot is superimposed on kraft paper to make a dense heavy-duty paper. Anaglypta is moulded wet to produce highly profiled embossing. Before application the voids created by the high relief should be filled with a paste of sawdust and paper adhesive.

(e) *Washable.* These are either varnished paper, plastic coated paper or complete plastics of vinyl or polystyrene. They are ideal in bathrooms where normal papers could decompose in the damp atmosphere.

Textiles

These are an alternative wallcovering of hessian, linen, silk, synthetic and natural fibres. Manufacturers set their own dimensional standards, therefore widths vary and lengths may be up to 50 m long. Suedes and felts are easily dyed and provide an attractive, smooth and richly coloured finish to prestige offices and living areas.

Paint (internal and external applications)

The purpose of paint is principally to protect and preserve the underlying material from decay or corrosion. Other functions are decoration by improving appearance, and hygiene by obliterating surface defects and porosity. The paint process is in three stages:

1 *Primer.* Adheres to the background and evens out surface porosity. With ferrous metals, controls rust.
2 *Undercoat.* Adheres to the primer, builds up the paint thickness and obliterates surface irregularities.
3 *Finish.* Adheres to the undercoat and provides a protective layer, colour and surface texture.

If a paint system is to function correctly, the combination of primer, undercoat and finish must be carefully selected for the specific background or surface.

Composition. Paint is a blend of the five components listed below. Proportions vary between manufacturers, and this accounts for quality and price differences.

1   *Pigment.* The solid component in paint responsible for colour and 'body'; e.g.:
    metallic oxides – red lead,
    coloured earths – yellow ochre,
    chemical compounds – Prussian blue.
2   *Binder.* Also known as the vehicle. This is an oil or varnish which binds together the other components and holds them in suspension.
3   *Drier.* Accelerates early drying by absorbing oxygen from the air and converting to a solid by oxidation. They are soluble metal compounds, e.g. lead or manganese in linseed oil or white spirit.
4   *Thinner.* A solvent which increases workability and penetration. Paint should be sufficiently thinned by the manufacturer, but application may justify additional thinning on site:
    Emulsion paints – water
    Oil paints – white spirit or turpentine.

5   *Extender.* Very finely ground material such as china clay. This is transparent in oil and has the function of increasing the adhesive qualities and hardness in a paint.

Specifications. Careful preparation of surfaces is necessary before the painting of them can begin.
1. *Wood.* Must be dry, ideally with a moisture content between 12 and 18%.

* Clean and smooth surface with a fine grade glasspaper.
* Dust surface and apply shellac knotting compound to seal all knots.
* Prime surface with white or pink lead primer, or a leadless white primer if on furniture, toys etc. Acrylic primer is also acceptable.
* Lightly sand primer and fill cracks and fissures with stopping compound.

- Apply two layers of undercoat and one or two finishing coats for internal and external use respectively.
- For a high quality finish, lightly sand the surface between application of each paint layer.

2. *Plaster and cement based surfaces.* Completely dry the surface.

- *Emulsions.* Apply in sufficient layers to obliterate surface irregularities and porosity differences.
- *Gloss.* Apply alkali-resisting primer, as alkalis can react with oil-based paints to cause tackiness. Follow with two coats of undercoat and one or two finish coats as necessary.

3. *Metalwork.* Clean off dirt, grease and loose scale. Wire brush rusty areas.

- Patch prime with rust inhibitor if rusty areas exist.
- Prime all surfaces with red lead or metal chromate primer.
- Apply two oil-based undercoats and at least one finishing coat.

Paint application. The majority of surfaces are treated by brush or cylinder roller. Both are shown in Fig. 11.13 with brush variations for special application or surfaces. The enlarged detail of a flat brush shows a hardwood or plastic handle attached to the filling by a chrome or nickel plated stock riveted to both. The

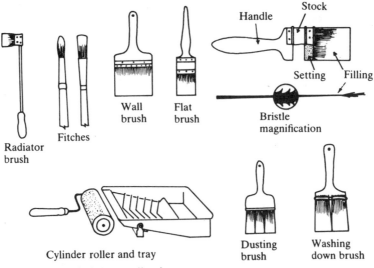

Fig. 11.13 Paint application

setting is an epoxy resin adhesive used to secure the brush filaments together.

Traditional filament is hog's hair or bristle, shown in magnified detail with natural serrations which hold the paint, and a long taper and end flagging for controllable application. Horse mane or tail are cheaper natural alternatives but these lack the strength, taper and end flagging of bristle. Glass and nylon fibres are synthetic imitations having excellent wear but poorer holding and finishing properties. Some manufacturers combine natural bristle with nylon, to compromise the cost of bristle with the good wearing qualities of nylon.

Cylinder rollers are particularly useful for emulsion painting large wall or ceiling areas. Savings in labour time are considerable and various textured finishes are possible by using roller covers of mohair, lambswool, sponge or nylon filaments.

External finishes

Ground

Clay and concrete bricks

These are now very fashionable for garage forecourts, car parks, access drives and patios. Conventional standard good quality building bricks (e.g. stocks) may be used, laid in a variety of patterns as may be seen from Fig. 11.14. These are generally only used for garden features and the bedding, which should be flexible to absorb climatic effects, need only contain about 50 mm of crushed rubble or similar granular filling, superimposed with sharp sand or ashes as shown in Fig. 11.15. Bricks are laid and levelled with a spirit level, between timber or other suitable edging and the 10 mm joints brush filled with a dry weak cement and sand mortar in the ratio of 1:5.

Concrete floor blocks are specifically manufactured for the heavy loading expected from vehicular traffic. Various colours and shapes are made, as may be seen in Fig. 11.16. Also shown is an edge sectional detail using rigid restraint from conventional kerbs and channels. Bedding is similar to brick flooring in gardens, except the flexible sub-base of crushed chalk, rock, slag or fuel ash must be at least 75 mm deep with a 50 mm minimum sharp sand laying course to accommodate greater loads. Roads are often produced on a sub-base of 150 to 200 mm thick lean mix concrete

Fig. 11.14 Brick flooring, pattern variations

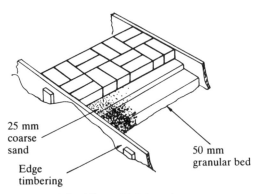

25 mm
coarse
sand

Edge
timbering

50 mm
granular bed

Fig. 11.15 Bedding of brick paths

composed of 20 to 40 mm aggregate to provide a compressive strength of about 10 N/mm^2.

Concrete paving slabs

Concrete paving slabs or flagstones are composed of cement and igneous rock aggregate. They are made in 50 mm or 63 mm thicknesses for pedestrian and vehicular access respectively.

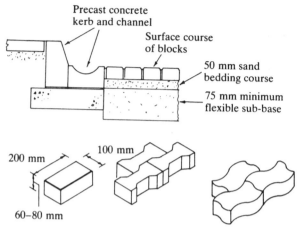

Fig. 11.16 Concrete block flooring

Table 11.4

BS type	Size (mm)
A	600 × 450
B	600 × 600
C	600 × 750
D	600 × 900

BS 368 provides four size classifications, represented in table 11.4.

Paving slabs are usually laid in regular formation as shown in Fig. 11.17. Sometimes broken slabs are used for garden paths and drives as a crazy paving feature. Whichever is used, the preparation after stripping the topsoil will be a 75 mm minimum flexible sub-base and a 25 mm dry cement and sand (1:3) mortar bedding with firm edge restraint from kerbing or purpose-made precast concrete edging. Joints of between 6 mm and 10 mm are left between slabs to be brush filled with dry mortar to match the bedding. Where used for vehicular access the sub-base should be at least 75 mm thick concrete with a cement/aggregate composition of 1:2:4.

Wall

The following section considers briefly, rendered, tiled and timber boarded finishes.

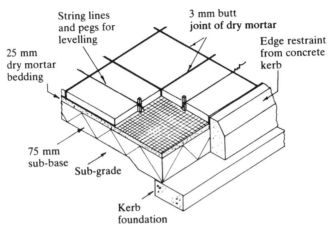

25 mm dry mortar bedding

String lines and pegs for levelling

3 mm butt joint of dry mortar

Edge restraint from concrete kerb

75 mm sub-base

Sub-grade

Kerb foundation

Fig. 11.17 Concrete paving flags

Rendering

Rendering is traditionally a plastering trade, using cement, lime and sand aggregates mixed with water to a workable consistency. For general use, mix ratios between 1:½:4 to 1:1:5 or 6 parts sharp sand will suit backgrounds ranging from expanded metal lathing to dense brick or block. Joints should be raked out 12–15 mm to provide a key for the first coat, or the surface adhesion may be improved by application of wire mesh as shown in Fig. 11.18, or a special lathing to timber framed walls as shown in Fig. 11.19. If the surface is particularly smooth and dense, it should be hacked to

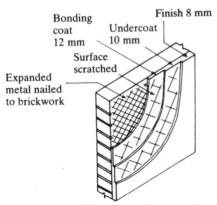

Bonding coat 12 mm

Undercoat 10 mm

Finish 8 mm

Surface scratched

Expanded metal nailed to brickwork

Fig. 11.18 Rendering a difficult surface

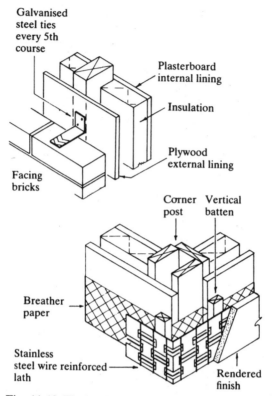

Galvanised
steel ties
every 5th
course

Plasterboard
internal lining

Insulation

Plywood
external lining

Facing
bricks

Corner Vertical
post batten

Breather
paper

Stainless
steel wire reinforced
lath

Rendered
finish

Fig. 11.19 Timber framed wall construction and finishing

provide a key or spatterdashed. This is strong cement and sand
(1:2) mixed to a slurry, trowelled thinly to the surface to leave a
rough surface.

The number of coats will depend upon the degree of exposure
and uniformity of the background. One 10 mm layer is adequate
where the background is regular and sheltered. Elsewhere the
brick or block joints are likely to 'grin' through in damp weather,
and two or possibly three coats are necessary.

Finishes may be smooth, textured, roughcast or pebble dash.
Smooth finishes are obtained by using a fine sand aggregate and
finishing with a steel trowel. Unless drying can be carefully
controlled, this finish has a tendency to crack and discolour.
Textured finishes are obtained by treating the final coat with a
brush or toothed implement to create decorative patterns in the
wet surface. These look attractive initially, but tend to attract dirt

Table 11.5 Render application

Application	Thickness (mm)
Undercoat	8–13
Spatterdash	8–16
Smooth finish	6–10
Textural finish	10–13 (3 mm surface treated)
Pebble dash	10

in the recesses. Roughcast is an irregular finish achieved by throwing the finish coat onto the wall. Some trowelling is necessary to regulate the effect. Pebble dash is obtained by throwing small pebbles onto a strong mortar finish coat. It has very good weathering qualities, provided the binding coat does not crack. Table 11.5 is provided as a guide to the thickness of the different applications mentioned.

Tile cladding

Plain tiling is an attractive alternative to brick cladding an external wall. Tiles are generally applied from first floor to eaves or as a gable feature. The background may be lightweight concrete blocks or a timber framework, both supporting 38 × 19 mm horizontal

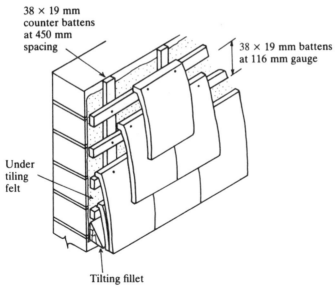

38 × 19 mm counter battens at 450 mm spacing

38 × 19 mm battens at 116 mm gauge

Under tiling felt

Tilting fillet

Fig. 11.20 Tile hanging to brick or block walls

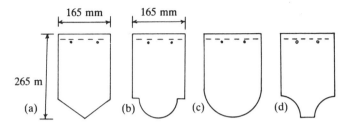

Fig. 11.21 Feature tiles (a) Arrowhead (b) Club (c) Beaver tail
(d) Scalloped

battens at 116 mm spacing or gauge. This corresponds with a lap of 32 mm – the amount that each tile overlaps the tile but one below. Every tile is nailed, and counter-battening shown in Fig. 11.20 is advisable to correct surface irregularities and to create an air void to improve thermal insulation. Undertiling felt is an additional waterproofing layer, advisable on porous concrete block backgrounds. Figure 11.21 shows some of the ornate tiles made in clay or concrete for tiled feature work to walls.

Timber cladding

Timber cladding of moulded softwood boards is also a popular

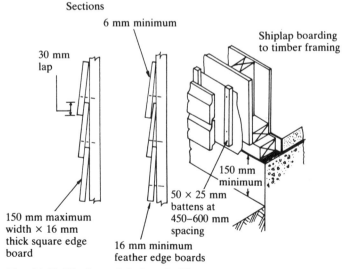

Fig. 11.22 Horizontal timber cladding

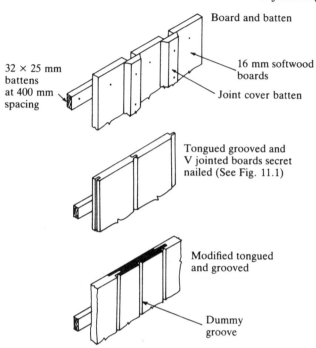

32 × 25 mm battens at 400 mm spacing

Board and batten

16 mm softwood boards

Joint cover batten

Tongued grooved and V jointed boards secret nailed (See Fig. 11.1)

Modified tongued and grooved

Dummy groove

Fig. 11.23 Vertical timber cladding

treatment to upper floor walls and gables. The extent permitted by Part B (fire) to the Building Regulations is restricted by position, location and size of the building. Additionally only durable materials are permitted which requires softwoods to be treated with approved preservatives. The Regulation allows timber boards of at least 16 mm thickness and featheredge boards 16 mm minimum at the thicker edge reducing to at least 6 mm at the thinner edge. These are shown in Fig. 11.22 with variations in tongued and grooved vertically fixed feature boarding detailed in Fig. 11.23.

12

Service installation

Water supply

Every habitable dwelling should have a piped supply of wholesome water controlled by a tap within the building. The pipework is installed with regard to waste prevention, contamination, damage from frost and corrosion; it should also have reasonable access for maintenance. These requirements are controlled by the local water authorities, who operate independently of local authority planning and building control departments. Each authority issues by-laws, most of which have been incorporated into nationally acceptable model by-laws. Figure 12.1 shows a typical supply service from the water authority's main to the householders stop valve within the dwelling. This incorporates provision for settlement of the main and contraction of the supply pipe, by snaking the branch pipe

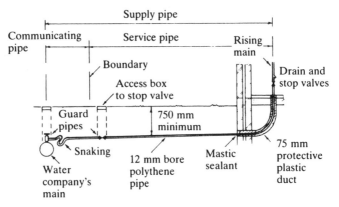

Fig. 12.1 Domestic water supply

close to the main. A minimum cover of 750 mm is shown as sufficient protection from frost and vegetation.

The water main has a stop valve located to its crown, to permit air to escape and to avoid loose rust deposits over the outlet. The communicating pipe links the supply with a stop valve inside the property boundary. This valve is not always provided, if it is convenient for the householder to have access to the valve on the main. Whether this valve is provided or not, the water authority is responsible for the repair and maintenance of the communicating pipe. The service pipe from the boundary to the isolating valve within the building, is the householder's responsibility, and must be maintained and repaired at his expense. Where this pipe enters the building it is protected from settlement by a clay or plastic pipe duct of 75 mm diameter. Alternatively it could enter below a small lintel, with the gas and electrical supplies. Each service is surrounded with a compressible packing to accommodate movement.

Materials for supply pipework have included lead, galvanised steel and copper. Most installations are now polythene to BS 6572 (Blue). This has the advantage of simple laying from coils, non-corrosive composition and flexibility. Jointing is by compression fittings with a copper ferrule located inside the pipe ends. This type of coupling is shown in Fig. 12.2, with non-manipulative and manipulative compression couplings for use on copper tube. Manipulative couplings are particularly specified for underground water supply pipes, as the swaged pipe ends prevent any joint movement and possible leakage.

Cold-water distribution

Requirements for distribution of cold water within a building differ between the local water authorities. In northern areas of Britain and other highland regions where water is trapped and processed at a relatively high level, it is possible for the water authorities to supply all water outlets with a direct constant pressure. Elsewhere the water is pumped to buildings or to high-level service reservoirs which gravity-feed the community. In these areas, each dwelling has a large cistern (230 litres capacity) to supply all water outlets except the kitchen tap, indirectly. This is designed to reduce pumped supply costs during peak demand periods. Table 12.1 compares both direct and indirect supplies and Figs 12.3 and 12.4 show respective systems for installation in houses.

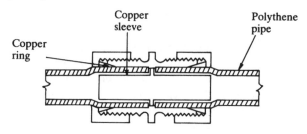

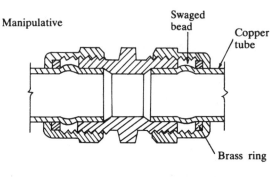

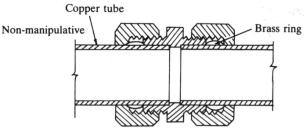

Fig. 12.2 Compression joints

Table 12.1 Comparison of direct and indirect water supply

Direct	Indirect
Less pipework	Higher installation costs.
Smaller capacity cistern (115 litres)	Cistern capacity 230 litres, therefore adequate emergency supply.
Drinking water at all outlets	Less pressure on taps and valves, therefore less wear and maintenance.
Limited emergency supply	Constant supply pressure.
Variable outlet pressures	Less demand on water main.

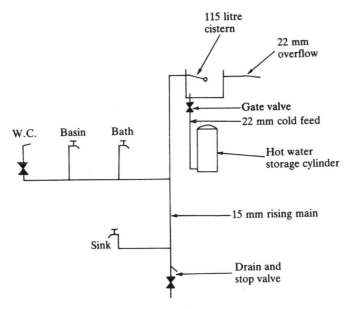

Fig. 12.3 Direct cold water supply

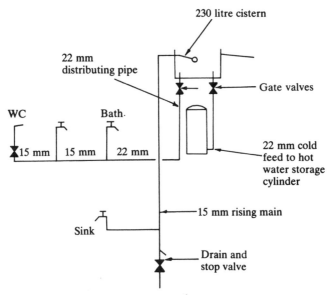

Fig. 12.4 Indirect cold water supply

Cisterns and ballvalves

Cistern capacity is specified by the amount of water contained under the control of a ballvalve, which differs considerably from the capacity to the rim as may be seen in Fig. 12.5. Modern cisterns are manufactured in polypropylene to the requirements of BS 4213. Ballvalves are an automatic water-controlling device, containing a valve which functions in response to the level of a plastic ball attached to the brass lever arm. Figure 12.6 shows sectional details of the BS 1212 ballvalves. The piston valve has a long history, including a 'Portsmouth' variation which has a non-detachable body and brass seating. This has to be reground if it wears, whereas the illustrated type is plastic and may be replaced. Diaphragm ballvalves are a relatively recent innovation designed by the Building Research Establishment. The advantages are less moving parts in contact with the water; therefore less erosion and less chance of 'water-hammer' – vibration through the pipework.

Control valves

Water control of simple installations is with either stop valves or gate valves. The difference is shown in sectional detail in Fig. 12.7.

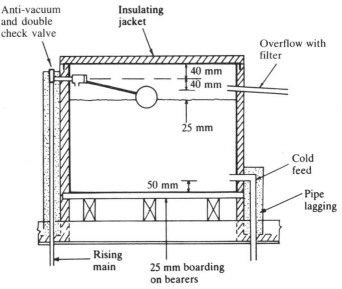

Fig. 12.5 Installation of cold water storage cistern

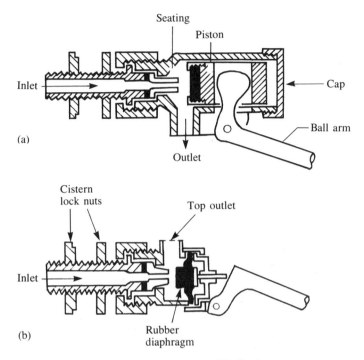

Fig. 12.6 Ballvalves (a) piston type, (b) diaphragm type

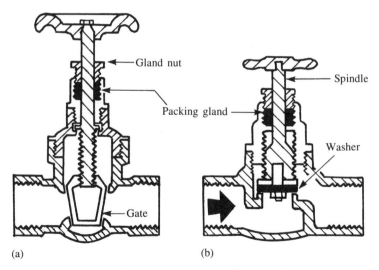

Fig. 12.7 Pipeline valves (a) gate valve, (b) stop valve

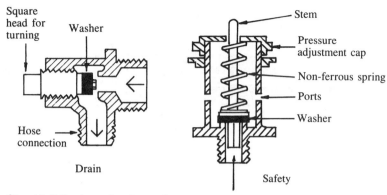

Fig. 12.8 Drain and safety valves

The stop valve is physically similar to a standard tap, with a renewable washer which is raised or lowered onto a seating by rotation of the handle. They are used on all water main controls as the slow operation and absorbency of the washer reduce the possibility of impact and water hammer through the pipework. Gate valves are a cheaper device which again operate by several rotations of a handle. The gate is a metallic sluice which slides up and down and will wear with use and fail to seat properly. Under high pressure the sluice could vibrate, which coupled with its poor seating properties prohibit its use on mains-pressure supplies. Therefore it is only used as an economy measure on low-pressure cistern supplies. Figure 12.8 shows a drain valve for use at the base of the rising main or other low level drain points, and a safety valve used to release excessive pressure on hot-water installations.

Hot water

Hot-water supplies originate from a centralised heat source. Historically this has been a solid-fuelled back boiler behind an open fire or free-standing stove. Modern hot-water supplies are centralised from gas- or oil-fired boilers which combine both hot water and central-heating functions. Occasionally, solid fuel is employed in modern systems where there are gas and oil supply difficulties. Supplementary heat energy may be obtained from an electric immersion heater in the hot-water storage cylinder.

Hot-water system components are principally boiler, hot-water storage cylinder and cold-water storage cistern. An expansion cistern will also be necessary where the cylinder water is heated indirectly. Figures 12.9 and 12.10 show possible pipework and

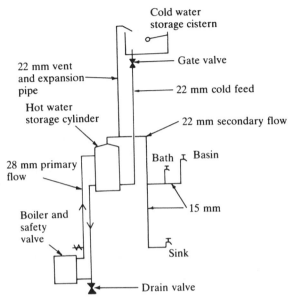

Fig. 12.9 Direct hot water system

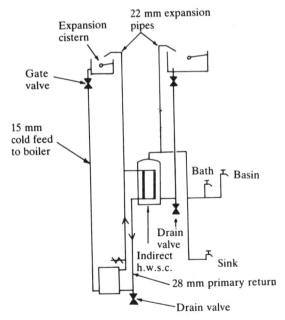

Fig. 12.10 Indirect hot water system

component location for simple domestic hot-water systems. In both, the cylinder on the first floor is shown as close as possible above the ground floor located boiler. This is to encourage gravity circulation of water by density change or convection currents in the primary flow and return pipework.

Direct system

The cold-water storage cistern supplies both the hot water storage cylinder and the boiler. Consequently, water heated in the boiler circulates to the cylinder, is stored until required, and is replaced with fresh water from the cistern. In hard water areas (underground water sources, in chalk or limestone sub-strata) the boiler and associated pipework become lined with calcium. This 'furring' can eventually block the pipework and cause damage from boiler explosion. Although the installation of a direct system is relatively cheap, it must only be used in soft water areas. Also, it is unsuited to radiator installations, as water drawn off for a bath, etc. will be drawn through the cylinder and the radiators, noticeably affecting their heat output.

Indirect system

This system employs a feed and expansion cistern in addition to the boiler, cylinder and storage cistern. The expansion cistern has a very small capacity of about 36 litres, sufficient to provide a filling and topping up function if required. The pipework is contained in two separate circulation systems. Firstly, the boiler heats and circulates water through a pipe coil inside the storage cylinder and then back to the boiler. By this means, stored water in the cylinder is heated indirectly and the water in the boiler primary pipework and coil is never drawn off. Fresh water containing calcium cannot penetrate the boiler, and 'furring-up' is impossible as the original water from the expansion cistern is sealed in. The second circulatory system is simply a cold supply to the cylinder from the storage cistern. This is heated indirectly from the coil, drawn off at taps and replenished as necessary. The heat exchange from the coil is insufficient to cause serious 'furring', but constant use of an immersion heater will cause sufficient scaling to limit its efficiency.

Simple heating circuits

Radiators are the most popular and economic heat source for small buildings. They are made from pressed steel in single, double or triple panels, and some have additional finning to the back to increase heat output. Figure 12.11 shows a standard radiator with four threaded tappings and typical attachments.

Systems

Radiators will only operate successfully if installation pipework is indirectly connected to the hot-water system, otherwise hot water and heating are interdependent and control of either affects the other. The simplest of systems using gravity or convection currents for hot-water circulation is shown in Fig. 12.12. This has limited output, as in excess of two radiators would noticeably affect hot-water supply to the indirect cylinder.

Accelerated systems

Gravity circulation to radiators is slow and very dependent on precise alignment and position of pipe runs. This is unacceptable for modern heating standards and installation procedure. To achieve a rapid heat response by accelerating the water flow rate, a circulator is positioned in a separate heating pipe circuit to the hot water primary flow and return. Artificial circulation provides several advantages:

1 Simplified pipe runs.
2 Small diameter pipes.
3 Fuel economy, as heating return water temperature is higher.

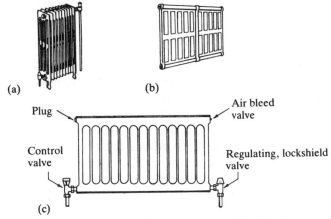

Fig. 12.11 Radiators (a) Cast iron column (b) Cast iron panel (c) Pressed steel

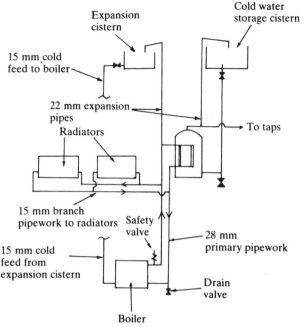

Fig. 12.12 Simple gravity hot water and heating system

Gravity flow 80 °C, return 60 °C.
Accelerated flow 80 °C, return 70 °C.

4   Higher average temperature, therefore greater heat emission from radiators.

One- and two-pipe accelerated systems

These are shown in Figs 12.13 and 12.14 respectively. The one-pipe system economises in pipework, but operating efficiency is limited because control of one radiator affects the balance of others and hot water which has cooled in each radiator returns to the heating main to supply following radiators with cooler water. The last (index) radiator receives only tepid water.

Drainage

Pipes

Drainage installations are now nearly always undertaken in clayware or plastic pipes. Clay pipes and fittings are comparably cheaper than plastic, but the shorter pipe lengths and more precise

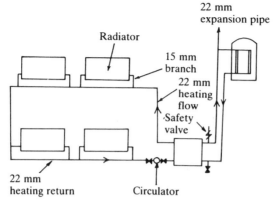

Fig. 12.13 One-pipe heating system

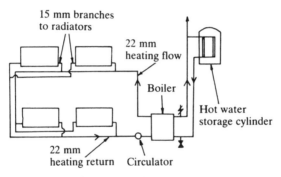

Fig. 12.14 Two-pipe heating system

bedding necessary with this rigid material are responsible for higher laying costs. Plastic conversely, is expensive to purchase but with pipe lengths up to 9 m, jointing and laying time considerably reduced.

Drainage pipework serving dwellings and other small buildings ranges between 75 mm nominal bore (rainwater), 100 mm (foulwater) and 300 mm nominal bore at the upper extreme. Pipe diameters in excess of 300 mm are normally part of the civil engineering contract which relates to main sewers below road surface level.

Clayware — BS EN295 (Nominal bore — 100, 150, 225 and 300 mm)

Traditional jointing and bedding. The traditional clay drain pipe is manufactured in 300 mm to 3 m effective lengths with spigot and socket connections shown in Fig. 12.15. Tarred gaskin

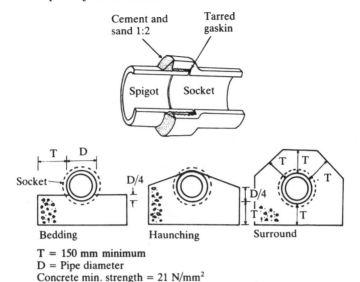

T = 150 mm minimum
D = Pipe diameter
Concrete min. strength = 21 N/mm^2

Fig. 12.15 Clayware drainpipe – traditional jointing and bedding

is used to centralise the spigot in the socket and the joint is sealed with a strong cement and sand mortar of 1:2 ratio. Where simply laid in the trench bottom and backfilled with excavated soil, ground movement and building settlement have been responsible for many pipe fractures. BS 8301 recommends effective concrete support for these pipes and this is also shown in Fig. 12.15. However, for success the concrete strength is critical and many poorly supervised installations, where the concrete is made very wet to aid placing, have subsequently led to pipe failures. Also a high water table or a rain-flooded trench will ruin the concrete strength.

Flexible jointing and bedding. The problem of pipe barrel failures and an awareness of ground movement potential led to the development of a modified spigot and socket joint. This is shown in Fig. 12.16 containing polyester moulded linings to both spigot and socket and a rubber 'O' ring fitted to the recess left in the spigot moulding. This permits sufficient angular movement and lineal draw without leakage, provided the pipes are bedded in a flexible medium. Nominal bore sizes up to 300 mm are 150, 225 and 300 mm in effective lengths of 1.5 m.

An alternative for 100 and 150 mm nominal bore pipes with plain ends in effective lengths of 1.6 m is also shown in Fig. 12.16.

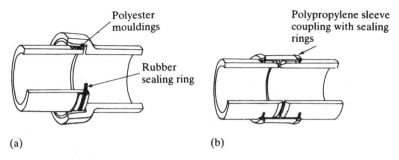

Fig. 12.16 Clayware drainpipe – alternative jointing (a) Spigot and socket (b) Plain end pipe

Jointing is with a polypropylene sleeve coupling, containing a sealing ring inside the rim at each end. Again a flexible bedding medium is essential and a comparison of techniques is shown in Fig. 12.17. Table 12.2 specifies the bedding material and effectiveness as a bedding factor. This is the ratio of vertical loading the pipe will carry in comparison to the British Standard test load, i.e. class S is 2.2 times better.

Table 12.2 Comparison of pipe bedding techniques

Class	Material and technique	Bedding factor
A	Reinforced concrete cradle*	3.4
A	Plain concrete cradle*	2.6
S	360° granular surround	2.2
B	180° granular support	1.9
Fss	Flat layer, single size granules	1.5
N	Flat layer, all in aggregate	1.1
D	Natural trench bottom	1.1

* 12 mm wide vertical gaps formed in the concrete with fibre board or expanded polystyrene, at 5 m maximum interval.

Unplasticised polyvinyl chloride (uPVC) BS 4600 and BS 5481

BS outside diameters; 110 and 160 mm
Non-standard diameters; 82, 200, 250 and 315 mm
Standard lengths 1, 3 and 6 m
Non-standard lengths 2 and 9 m.

Jointing. Jointing of uPVC pipes is by solvent-welded spigot and sockets or push-fit connection with sealing rings. Solvent

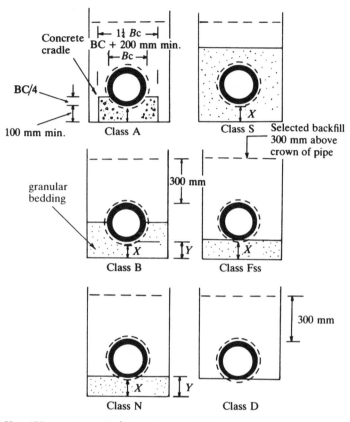

Y = 100 mm min. or Bc/6 in uniform soil (50 mm for plain end pipes with
 sleeve couplings.
Y = 200 mm min. or Bc/4 in rock or mixed soils containing rock, etc.
 (150 mm for plain end pipes with sleeve couplings).
X = 50 mm min. under sockets in uniform soil.
X = 150 mm min. under sockets in rock or mixed soils containing rock, etc.

Fig. 12.17 Flexible bedding techniques

welds are only advisable for fabrication of fittings and inspection
chamber connections, as they fail to absorb the thermal movement
expected of this pipe material. Push-fit connections with sealing
rings are manufactured in conventional spigot and socket form as
shown in Fig. 12.18 and as a double socket or collar, also shown in
Fig. 12.18.

 Bedding. uPVC pipes are sufficiently flexible to absorb
modest ground movement, therefore bedding is not as critical as

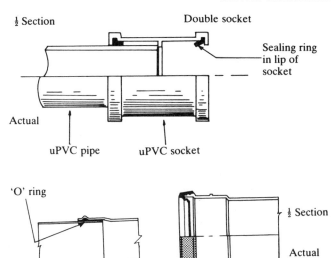

Fig. 12.18 Jointing of uPVC drainpipes

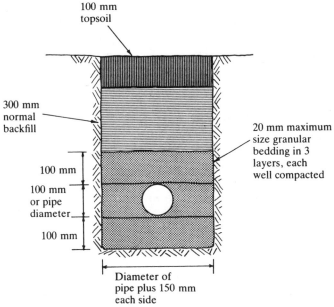

Fig. 12.19 Bedding and backfilling plastic drainpipes

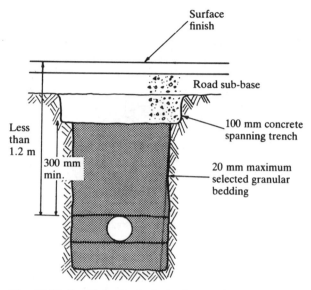

Fig. 12.20 Bedding plastic drainpipes under roads

with rigid clayware. Figure 12.19 shows acceptable bedding from 1.2 to 6.1 m depth to the pipe crown which will give adequate protection against deformity. Depths within 1.2 m are acceptable if the surface is garden or footpath, but if below a road or driveway a concrete protecting raft as shown in Fig. 12.20 must be provided. Alternatively, and for pipes over 6.1 m deep, a complete surround of concrete (see Fig. 12.15) is acceptable. Concrete is not harmful to uPVC, but it is difficult to apply as the pipe floats unless filled with water. Also, joints should be wrapped in polythene to prevent the penetration of cement.

Systems of drainage

Conveyance of surface water and foul water from small buildings is by combined or separated pipework. The combined system is the simplest and most economic to install, but as both surface water and foul water share the same drain and sewer, the cost of sewage treatment is high due to the additional volume of surface water. Consequently, very few local authorities will accept this system, even though the drains and sewers benefit from a thorough washing during inclement weather.

The separate system doubles the drainage pipework necessary to serve a building. Surface water is collected in a sewer below

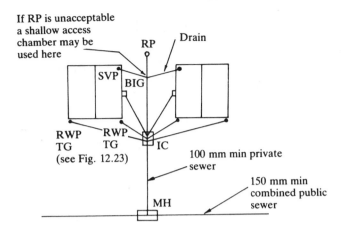

If RP is unacceptable
a shallow access
chamber may be
used here

RP   Drain

SVP
BIG

RWP   RWP
TG   TG   IC
(see Fig. 12.23)   100 mm min private
                        sewer

                            150 mm min
                            combined public
                            sewer

MH

Alternative

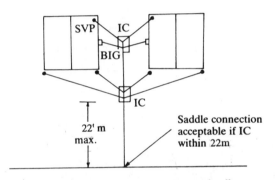

SVP   IC

BIG

IC

22' m   Saddle connection
max.   acceptable if IC
                     within 22m

Key RP = Rodding point   TG = Trapped gulley
      SVP = Soil and vent pipe   SAC = Shallow access chamber
      BIG = Back inlet gulley   IC = Inspection chamber (up to 1m)
      RWP = Rainwater pipe   MH = Manhole (over 1m deep)

Fig. 12.21 Combined drainage systems

road level for transfer to a stream, river or other convenient
outlet. Figures 12.21 and 12.22 show some applications of both
systems and Fig. 12.23 facilities for connecting rainwater pipes to
drains. The size of drains acceptable for up to 20 dwellings is
100 mm laid to a minimum fall of 1 in 80. Under roads, the local
authorities usually insist on at least a 150 mm diameter pipe to
accommodate future development. This size of pipe has sufficient capacity for at least 100 dwellings with a minimum fall of
1 in 150.

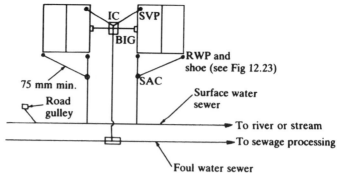

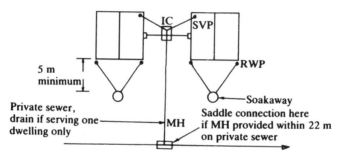

Fig. 12.22 Separate drainage systems

Soakaways

In granular subsoils with good drainage properties, soakaways may be constructed. The simplest type is a pit in the ground, filled to within about 200–300 mm of the surface, with a good draining medium such a brick rubble. A polythene sheet overlays the rubble to prevent topsoil clogging the drainage gaps. Figure 12.24 illustrates this type of soakaway for use with a single building and Fig. 12.25 shows a larger version, built from precast perforated concrete rings suitable for several dwellings or large buildings. Soil percolation tests and graphical references may be used for calculating soakaway capacity, or this simple empirical formula which gives adequate guidance for small buildings:

$$C = \frac{AR}{3}$$

where C = Capacity in m^3
      A = Area of roof or surface to be drained (on plan) in m^2
      R = Rainfall, maximum expected in m/h

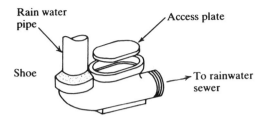

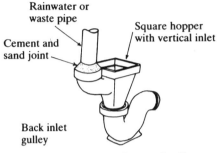

Fig. 12.23 Rainwater shoe and gulley

D = depth below pipe invert level, approximately the same as *d*, the diameter.

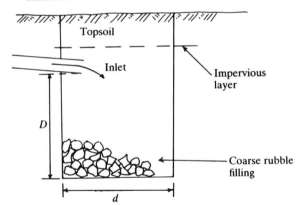

Fig. 12.24 Simple rubble and filled soakaway

For example:

Roof area = 100 m²
Maximum rainfall = 30 mm/h (0.03 m/h)

$$C = \frac{100 \times 0.03}{3} = 1 \text{ m}^3$$

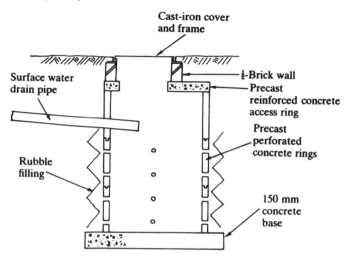

Fig. 12.25 Perforated soakaway

Soakaways must be at least 3 m from buildings to prevent undermining of foundations. Most local authorities insist on 5 m.

Inspection chambers and manholes

Both inspection chambers and manholes are means for obtaining access to drains. The term manhole applies to all chambers which allow access for inspection, testing and clearance of obstructions, and the term inspection chamber is restricted to those having a maximum invert depth of 1 m.

Access should be provided at the following situations:

1   Every change in direction.
2   Every change in gradient.
3   At all drain junctions, except;
4   22 m maximum from the junction with a public sewer.
5   At 90 m maximum intervals on a straight drain run.
6   At the head of each drain length.

An inspection chamber at No. 6 may be substituted with a rodding point (see Fig. 12.21) if the local authority are agreeable. Provision is shown in Fig. 12.26.

Construction of manholes and inspection chambers is traditionally in brick and concrete as shown in Fig. 12.27. This form of construction is still appropriate for deep chambers in

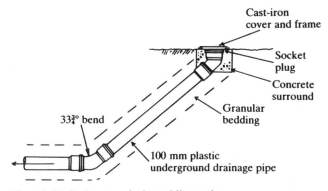

Cast-iron
cover and frame

Socket
plug

Concrete
surround

Granular
bedding

33¾° bend

100 mm plastic
underground drainage pipe

Fig. 12.26 Rainwater drain rodding point

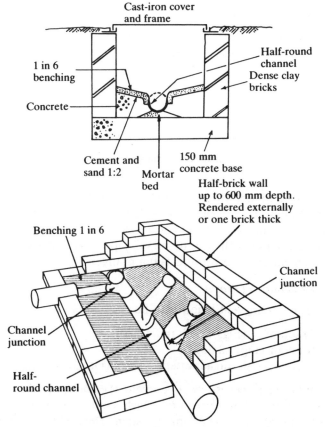

Cast-iron cover
and frame

1 in 6
benching

Half-round
channel

Dense clay
bricks

Concrete

Cement and
sand 1:2

Mortar
bed

150 mm
concrete base

Half-brick wall
up to 600 mm depth.
Rendered externally
or one brick thick

Benching 1 in 6

Channel
junction

Channel
junction

Half-
round channel

Fig. 12.27 Traditional brick inspection chamber

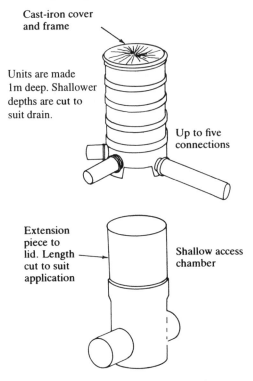

Cast-iron cover
and frame

Units are made
1m deep. Shallower
depths are cut to
suit drain.

Up to five
connections

Extension
piece to
lid. Length
cut to suit
application

Shallow access
chamber

Fig. 12.28 Plastic inspection chamber

excess of 1 m depth to invert (lowest level at which water will flow). Below this depth, most local authorities will accept alternative materials, and Fig. 12.28 shows a polypropylene chamber for use with clay or uPVC pipes and a plastic shallow access chamber for use up to 600 mm invert level.

Ventilàtion of drains

Systems of drain ventilation may be:

1   Traditional (Figs 12.29 and 12.30)
2   Modern (Fig. 12.31)
3   Modern with air admittance valve (Figs 12.32 and 12.35)

Ventilation is necessary to prevent the build-up of explosive gases and to relieve pressure which could interfere with the smooth operation of the drainage system. The traditional system is unnecessarily expensive, using high-level ventilation pipes to relieve sewer pressure and the soil and waste stack to ventilate each individual drain.

An interceptor trap separates each means of ventilation and a fresh air inlet is provided in the adjacent inspection chamber. Modern systems eliminate the interceptor trap and fresh air inlet, as each soil and waste stack successfully ventilates the sewer. An air admittance valve positioned above the stack flood level has the advantages of material savings plus no disruption to the roofing where the vent pipe would penetrate. Where this valve is used, every fifth house must have a conventional vent pipe to preserve atmospheric pressure in the drains and sewer. Conventional vent outlets must be at least 900 mm above the highest openable window or ventilator, and 3 m horizontally from such an opening.

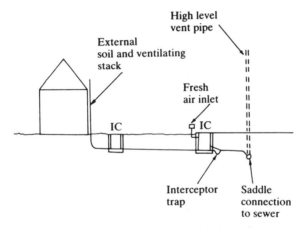

Fig. 12.29 Traditional ventilation of drain and sewer

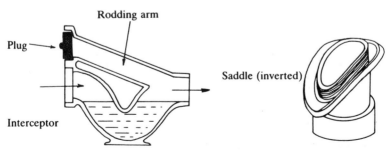

Fig. 12.30 Interceptor trap and saddle

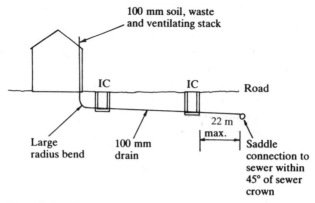

Fig. 12.31 Contemporary ventilation of drain and sewer

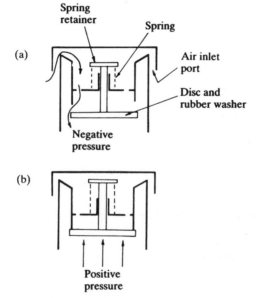

Fig. 12.32 Air admittance valve – operating principles (a) Open
(b) Closed

Sanitation

It is essential that all sanitary fittings, i.e. bath, basin, WC etc.,
connected to a drain are fitted with a water-filled trap. This is a

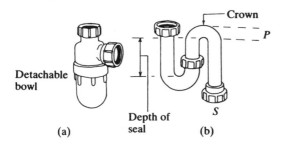

Fig. 12.33 Traps (a) Bottletrap (b) Conventional tubular trap

very simple double 'U' bend, shown in conventional form in Fig. 12.33. The lower bend traps a small volume of water, to prevent the foul air and gases from the drain and sewer penetrating a building. The bottle trap, also shown, is a tidier and more compact trap which functions in the same way. A secondary function of traps is to retain debris that might otherwise block the waste pipe. Access for cleaning is provided by detaching the bowl of a bottle trap or a section of the tubular trap. Traps attach to the screwed waste outlet on most appliances; the WC pan is an exception, with an integrated 50 mm water seal as shown in Fig. 12.34.

Loss of water seal

Poor design and installation of a sanitation system can cause the traps to lose their water seal. Bad workmanship or a faulty fitting could cause the trap to leak, but most seal loss problems are associated with induced siphonage or self siphonage. These are caused by branch pipes to appliances being too long, too small diameter or too steep in gradient. Compression or back pressure at the stack base is another source of seal displacement, caused by a small radius bend and a stack of insufficient diameter. The mechanics of these problems are shown in Fig. 12.34.

Single stack system

Prevention of trap seal loss was originally overcome by branch ventilating pipes attached to all wastes culminating in a ventilating stack. This is expensive to install and unpleasant in appearance, and has now been replaced by the single stack system of sanitation. This has one vertical stack to fulfil the discharge

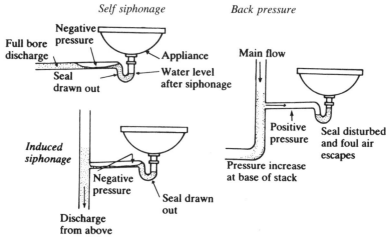

Fig. 12.34 Seal loss in traps

requirements from soil (WC) and waste (bath, basin and sink) fittings plus sufficient ventilation capacity. To perform correctly all branch pipes must have a minimal gradient (6° max) and length limited to the dimensions given in Fig. 12.35. This system is not restricted to houses, it may be used for high-rise buildings provided the stack diameter is increased accordingly.

Cesspools and septic tanks

Cesspools and septic tanks are acceptable methods for foul water containment and treatment where main sewers do not exist. A cesspool is simply an impervious container which is emptied periodically. A septic tank is a small-scale sewage treatment plant, which operates principally by the decomposition of solids by anaerobic bacterial activity in the absence of dissolved oxygen.

Both cesspools and septic tanks are produced in reinforced plastic materials for convenient location in prepared pit excavations. For purposes of illustrating the principles of operation, contemporary plastic with concrete construction is shown in Figs. 12.36 and 12.37.

Cesspool

Although cesspools are simple in construction, they have the disadvantage of requiring periodic emptying when full. This is costly, therefore a large capacity is desirable. For design purposes,

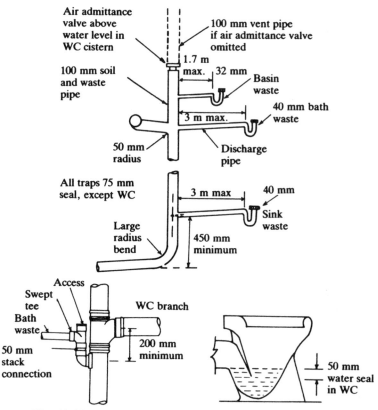

Air admittance valve above water level in WC cistern

100 mm vent pipe if air admittance valve omitted

1.7 m max.

100 mm soil and waste pipe

32 mm Basin waste

40 mm bath waste

3 m max.

50 mm radius

Discharge pipe

All traps 75 mm seal, except WC

3 m max

40 mm Sink waste

Large radius bend

450 mm minimum

Access

Swept tee

Bath waste

WC branch

50 mm stack connection

200 mm minimum

50 mm water seal in WC

Fig. 12.35 Single-stack sanitation

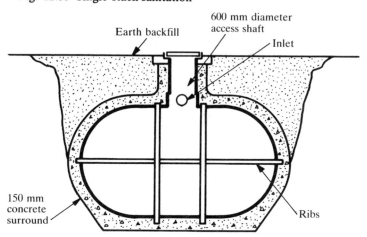

600 mm diameter access shaft

Earth backfill

Inlet

150 mm concrete surround

Ribs

Fig. 12.36 Cesspool

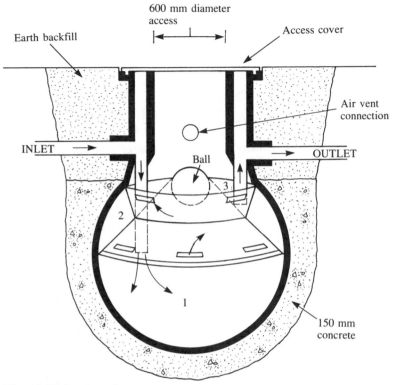

Fig. 12.37 Septic tank

the Building Regulations require a minimum capacity of 18 m³ (18 000 litres). At the other extreme, it is unlikely to be structurally viable to build a cesspool over 50 m³ capacity. Effective storage calculations may be based on the figure of 150 litres per head per day, with emptying taken as a minimum of 45-day intervals, e.g. 5-person household and desired emptying intervals of 45 days.

$$5 \times 150 \times 45 = 33\ 750 \text{ litres}$$
$$\text{or } 33.75 \text{ m}^3$$

Location is at least 15 m from a building with vehicular access no more than 30 m distance.

Septic tank

Periodic emptying is not so critical as with cesspools, but they should be desludged at 12-monthly intervals. Capacity is based on the simple formulae:

$$C = (180P + 2000)$$

where C = capacity in litres (min. 2720 litres)
P = number of persons normally using the tank (min. 4).

For example: 5 person household
$$C = (180 \times 5 + 2000)$$
$$= 900 + 2000$$
$$= 2900 \text{ litres or } 2.9 \text{ m}^3$$

Construction is in three separate compartments. The first compartment separates solids from liquids; heavier solids settle as sludge and lighter solids form a surface scum. The scum is a solid crust which excludes oxygen, thereby encouraging anaerobic action to reduce the volume of solids. The middle compartment is half the volume of the first and completes any further separation of solids that may be necessary. The outlet from the third compartment should feed a small biological filter of broken stone or rubble, where aerobic bacterial action will complete the process before the water is discharged. Discharge may be into a stream or river, or by percolation into underground porous strata by use of perforated land drain pipes. In either case permission will be required from the local water authority and they will probably examine treated water samples. Direct outflow from the settlement chambers into a water course without aerobic treatment is likely to be rejected by the water authority as this is merely a primary treatment; possibly of only one day's duration.

Oil and petrol interceptors

These chambers are essential where the surface water drains from garage forecourts, industrial premises, etc. discharge into a surface water sewer which has an outlet into a river or other water course. The construction may be in traditional brick and concrete or reinforced plastic, and is shown in principle in Fig. 12.38.

Function:
1 Polluted water enters compartment 1, and oil/petrol floats to the surface.
2 Cleaner water from compartment 1 rises up the dip pipe and is displaced into compartment 2.
3 The same process repeats from compartment 2 or 3 with smaller amounts of oil/petrol on the surface of compartment 2.

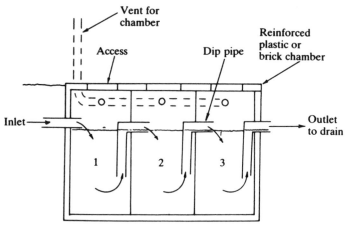

Fig. 12.38 Operating principles of a petrol interceptor

4 Compartment 3 contains virtually clean water and the outlet from this dip pipe discharges into the surface water sewer.
5 Surface oil and petrol are skimmed clean from each compartment at necessary intervals.

13

External works

Fencing

Fencing is for provision of security, screening or establishment of boundaries. Choice of materials vary; the most popular include timber, metal and concrete.

Timber

The style and quality of timber fencing ranges from the elementary chestnut paling wired to posts, to examples of very sophisticated joinery. Timber blends well with the environment and is naturally unobtrusive. If desired it can be painted or treated to preserve its durability with pressure-impregnated creosote or waterborne copper-chrome-arsenic solution. Painting is unsuccessful over creosote, but if waterborne preservatives are used and sufficient time is allowed for drying, paint decoration will be successful.

Examples of timber fencing

1. Chestnut pale and spile. Cleft chestnut pales are supplied in rolls, wired together with two or three lines of galvanised wire. They range in height from 900 to 1800 mm and receive support from 75 mm round chestnut poles at 1 m to 2 m spacing. Posts should be driven at least 500 mm into the ground as shown in Fig. 13.1 or set in concrete.

Chestnut spiles are stronger than the pales just described, and are driven at least 300 mm into the ground at approximately 150 mm spacing. Stout end and intermediate posts are again positioned at 1 m to 2 m spacing and connected to spiles with two lines of galvanised wire twisted alternately clockwise and anticlockwise between each spile.

2. Post and rail. This is an open barrier consisting of 150 ×

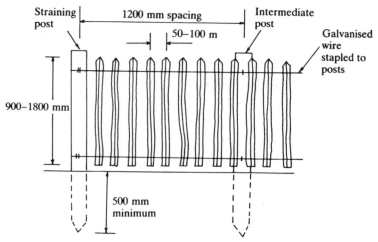

Straining post

1200 mm spacing

50–100 m

Intermediate post

Galvanised wire stapled to posts

900–1800 mm

500 mm minimum

Fig. 13.1 Chestnut pale fencing

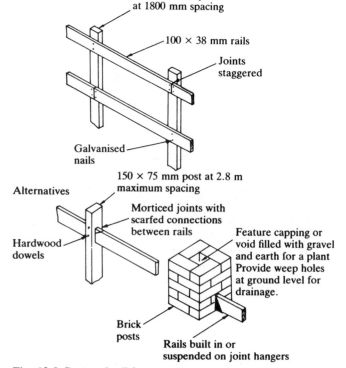

150 × 75 mm posts at 1800 mm spacing

100 × 38 mm rails

Joints staggered

Galvanised nails

150 × 75 mm post at 2.8 m maximum spacing

Alternatives

Morticed joints with scarfed connections between rails

Hardwood dowels

Feature capping or void filled with gravel and earth for a plant Provide weep holes at ground level for drainage.

Brick posts

Rails built in or suspended on joint hangers

Fig. 13.2 Post and rail fence

75 mm timber posts spaced at up to 2.8 m, and one to three horizontal rails of 100 × 38 mm section. Where connections between posts and rails are nailed, the posts should not exceed 1.8 m spacing and rails should nail to the longest face of the post. Wider post spacing may be used with mortice and tenon joints, provided the longest post face is used for the joint. Figure 13.2 shows the applications of these techniques.

3. Ranch. Ranch-style fencing is based on post and rail fence principles, except that the rails and intermediate gaps are of equal width as shown in Fig. 13.3. Also shown is secondary railing arranged to obscure the gap between the other rails. This is a feature which offers a degree of privacy and relief from wind pressure that close-boarded fences often suffer.

4. Palisades. Both post and rail, and ranch fences are frequently fitted with vertical boards or palings, shown in Fig. 13.4. The horizontal rails are now termed cant-rails, and they need to be

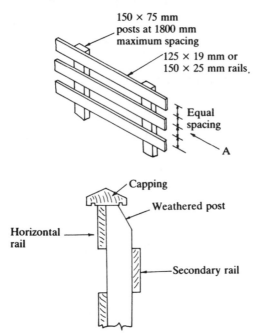

Elevation on A, with optional capping, weathered posts and secondary rails

Fig. 13.3 Ranch fencing

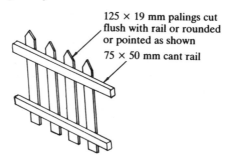

125 × 19 mm palings cut
flush with rail or rounded
or pointed as shown

75 × 50 mm cant rail

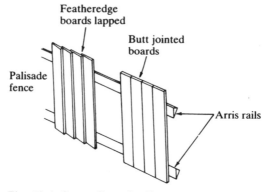

Featheredge
boards lapped

Butt jointed
boards

Palisade
fence

Arris rails

Fig. 13.4 Cant rails and palings

at least 50 mm thick to receive the paling nails. Rails may be the
triangular arris type also shown in Fig. 13.4 with the options of
close and feather-edge boarding. Arrises or rectangular rails are
morticed into rectangular posts.

5. *Close boarded.* These traditional style fences are very
durable and are used usually for height ranges from 1.5 m to
2.5 m. Posts are stout timber of 100 to 150 mm square section or
reinforced concrete. Both have two or three rectangular shaped
mortices to receive tenoned arrises which are hardwood dowelled
into timber posts and bolted to concrete. Overlapping
feather-edged boarding forms the pales, and this should terminate
about 150 mm above ground level for location of a gravel board.
This board will rot, but is more easily and cheaply replaced than a
series of rotting pales. Capping to posts and pales is an optional
protective feature as shown in Fig. 13.5.

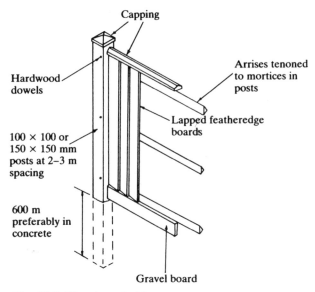

Capping

Hardwood
dowels

Arrises tenoned
to mortices in
posts

100 × 100 or
150 × 150 mm
posts at 2–3 m
spacing

Lapped featheredge
boards

600 m
preferably in
concrete

Gravel board

Fig. 13.5 Close boarded fence

6. Panels. Fencing in panels is convenient for quick site assembly, as these are supplied in factory made units of up to 1.8 m length and height. The most common panel comprises thin slats of wood interwoven between framed uprights. They are nailed to posts, which is of limited effect. The panels are not very robust, having a tendency to split if not creosoted at annual intervals. They are therefore unsuited to estate boundaries and other prominent positions, but are successful in restricted use for division of gardens and other minor boundaries. Figure 13.6 shows typical panel assembly.

7. Trellis. Trellis fencing is useful where high boundary fences are required but access for daylight is desired. They are usually factory made and often feature above a panelled fence to increase height without incurring potential wind damage. Trellis timber is normally about 32 × 19 mm in section and may be arranged at right-angles as shown in the prefabricated panel in Fig. 13.7. Also shown is a diagonal lattice arrangement, constructed *in situ*, attached to posts of at least 75 mm square.

All fencing described to a height of 1 m should have posts concreted at least 300 mm into the ground. Posts supporting

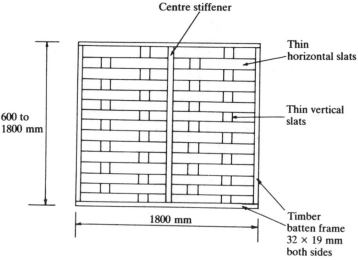

Fig. 13.6 Prefabricated interwoven panel

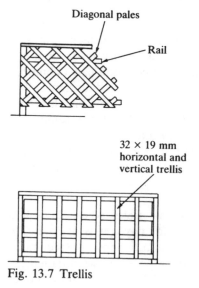

Fig. 13.7 Trellis

fences up to a height of 2.5 m should be concreted to a depth of 600 mm.

Metal

Metal fencing is divided into two categories: rigid or flexible. Rigid fencing is made up from stiff metal sections of round, square, flat or hollow profile. Flexible fencing includes mesh, wire and chain link materials supported by rigid posts of steel or concrete.

Rigid metal fencing

1. Continuous bar. This type of fence is frequently seen enclosing estates and farmland. It ranges in height from 1.05 to 1.35 m and is composed of flat or round section horizontal bars supported by flat, T or L-section posts at about 900 mm spacing. Figure 13.8 shows possible fixing of horizontal bars and posts.

2. Vertical bar (unclimbable). This type of fence has a flat, round or square top and bottom rail with round or square vertical bars at 115 to 150 mm spacing. Fence height ranges between 1.2 to 2.1 m with supporting posts at a maximum of 2.76 m. Posts should penetrate the ground 530 mm for fence heights between 1.2 and 1.35 m, 610 mm for fence heights between 1.5 and 1.8 mm and 760 mm for heights in excess of 1.8 m. Stay plates bolted to the posts at a position two thirds up should also be provided, and both stays and posts should be concreted in the ground. Figure 13.9 shows typical vertical bar fencing assembly.

3. Balustrade panels. Balustrade fencing is prefabricated

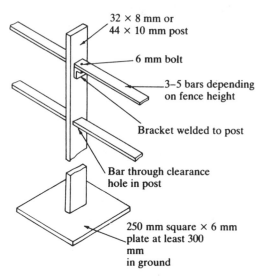

32 × 8 mm or
44 × 10 mm post

6 mm bolt

3–5 bars depending on fence height

Bracket welded to post

Bar through clearance hole in post

250 mm square × 6 mm plate at least 300 mm in ground

Fig. 13.8 Continuous bar fence

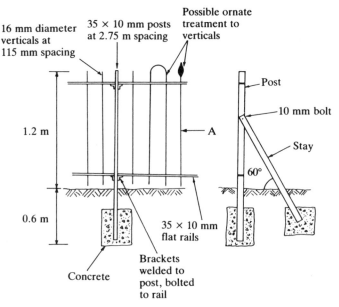

16 mm diameter verticals at 115 mm spacing

35 × 10 mm posts at 2.75 m spacing

Possible ornate treatment to verticals

Post

10 mm bolt

Stay

60°

1.2 m

A

0.6 m

35 × 10 mm flat rails

Concrete

Brackets welded to post, bolted to rail

Fig. 13.9 Unclimbable vertical bar fence

from mild steel rectangular hollow and flat sections. Security and boundary panels are fairly regular as shown in Fig. 13.10. They are manufactured in height ranges from 0.9 to 3 m for bolting to brick piers, concrete or steel posts. Figure 13.11 illustrates some elaborate panels forged from solid flat and square sections as supplementary featurework to brick walls or as individual fence panels bolted to posts. Panelling to brickwork is built-in directly or bolted to built-in brackets.

Flexible metal fencing

1. Chain link. Chain link fencing is particularly useful for enclosing sports grounds, e.g. tennis courts, as there is negligible light restriction with effective ball restraint. It arrives on site in rolls for heights ranging between 0.9 m and 3.6 m. Finish to the interlinked steel wire is by galvanising or plastic coating. Support is from concrete posts, or hollow or angle steel posts at 2 to 3 m spacing suspending two, three or four horizontal wire strands interlacing through the chain links. Fixing to posts is by tying with galvanised wire. As a security barrier, anti-intruder chain link fences are provided with cranked arm posts supporting two or three strands of barbed wire as shown in Fig. 13.12.

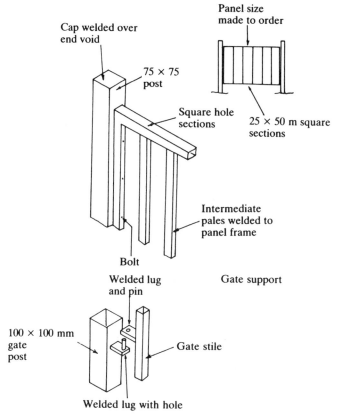

Fig. 13.10 Balustrade panelling

2. *Mesh.* Various fencing meshes are produced in expanded or perforated sheet form for attachment by wire to posts and suspended horizontal wires. Alternatively, panels with rigid steel bar edging are manufactured for direct bolt fixing to concrete or steel posts. Hexagonal wire mesh fencing more usually known as chicken or rabbit wire has smaller scale applications. It is available in rolls of various widths for stapling to timber posts or wiring to steel or concrete.

Concrete

There are many examples of concrete fencing, most following the layout and design of timber fencing, but offering a more robust structure. The appearance is less attractive, although some relief is obtained by varying surface textures and colours. All concrete

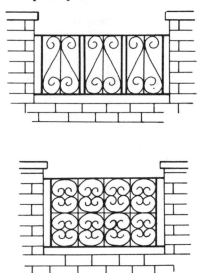

Fig. 13.11 Wrought iron feature panels

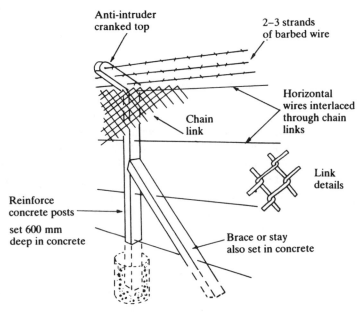

Fig. 13.12 Chain link fencing

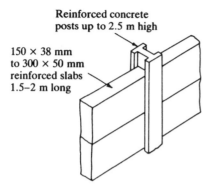

Reinforced concrete
posts up to 2.5 m high

150 × 38 mm
to 300 × 50 mm
reinforced slabs
1.5–2 m long

Fig. 13.13 Concrete post and panel

components are steel reinforced. The following two examples show simple concrete fences for use as garden or office boundaries.

1. Post and panel. Posts are grooved either side so that the section is similar to an I beam. Each groove or recess receives concrete slabs of 38 or 50 mm thickness × 150 to 300 mm width as shown in Fig. 13.13.

2. Picket or palisade. This is composed of integrally cast panels of pales and rails slotted into concrete posts as shown in Fig. 13.14.

Footpaths and drives

These may be constructed to obtain a flexible or rigid structure. Flexible construction of paving slab, brick and concrete block

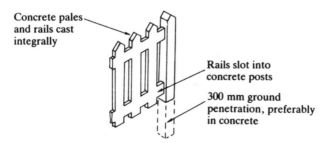

Concrete pales
and rails cast
integrally

Rails slot into
concrete posts

300 mm ground
penetration, preferably
in concrete

Fig. 13.14 Concrete picket or palisade

systems laid on a sand or ash bedding are considered in Chapter 11 under external finishes. Alternative flexible pavings and driveways may be constructed with a gravel or tarmacadam surface.

Gravel drives construction procedure is as follows:

1 Remove about 200 mm of topsoil.
2 Compact and consolidate the sub-grade.
3 Form concrete or brick edgings.
4 Lay and consolidate 100 mm of hardcore, (broken brick rubble).
5 Spread and level to about 50 mm thickness, 20 mm graded aggregate.

Note: Pea shingle tends to be too small, catching in car tyres and transferring to garage and other areas.

Figure 13.15 shows a typical section through this type of driveway. Tarmacadam surfacing to drives is expensive, therefore it is a false economy to try and make savings on the preparation. After removing sufficient top and subsoil, the following procedure should be applied:

1 Compact and consolidate the sub-grade.
2 Form edgings.
3 Lay and consolidate 100 mm of hard core.
4 Apply 75 mm layer of fairly dry, lean (1:15–20) concrete and compact.
5 Treat surface with bituminous emulsion or bitumenous 'tacking tar'.
6 Spread and level asphalt or bitumen-coated mineral chippings. (One layer is adequate for a footway, two minimum for a drive).

Figure 13.16 shows this construction in sectional detail, and also a section through the construction of a typical flexible estate road.

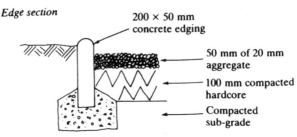

Edge section

200 × 50 mm concrete edging

50 mm of 20 mm aggregate

100 mm compacted hardcore

Compacted sub-grade

Fig. 13.15 Gravel driveway

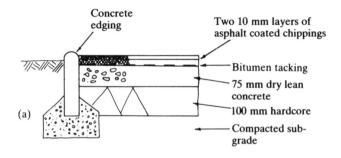

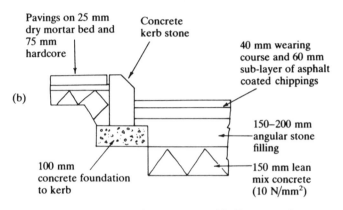

Fig. 13.16 Tarmac surfaces (a) Driveway (c) Estate road

Rigid roads

Rigid roads are manufactured in concrete primarily for estate access. As concrete is unable to absorb thermal movement and settlement without fracture, the surface appearance is spoilt by the provision of expansion, contraction and longitudinal construction joints. Figure 13.17 shows a typical plan arrangement of these joints for part of an estate road and Fig. 13.18 details the construction in section showing facilities for controlling cracking and absorption of movement. Dimensions given in Fig. 13.17 relate to a 125 mm mesh-reinforced concrete slab of 20 N/mm² compressive strength. For a slab thickness of 175 mm, which is more practical for contemporary traffic loading, contraction joints may be at 12.5 m intervals and expansion joints at 50 m intervals. Figure 13.19 shows possible kerb line construction including drainage through a precast concrete or clayware gulley.

Slab thickness 125 mm, concrete 20 N/mm²

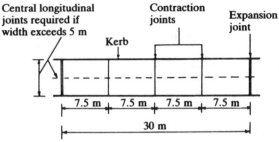

Central longitudinal joints required if width exceeds 5 m

Contraction joints

Expansion joint

Kerb

7.5 m | 7.5 m | 7.5 m | 7.5 m

30 m

Fig. 13.17 Concrete estate road plan

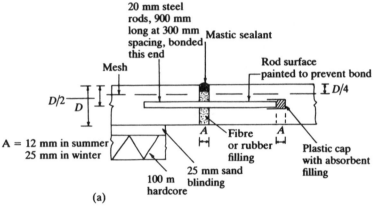

20 mm steel rods, 900 mm long at 300 mm spacing, bonded this end

Mastic sealant

Mesh

Rod surface painted to prevent bond

$D/4$

$D/2$ D

A Fibre A or rubber filling

Plastic cap with absorbent filling

A = 12 mm in summer 25 mm in winter

100 m hardcore

25 mm sand blinding

(a)

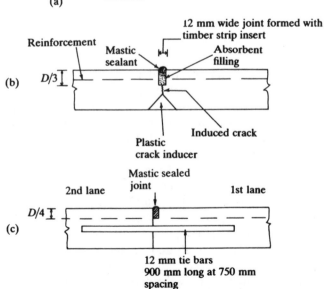

12 mm wide joint formed with timber strip insert

Reinforcement

Mastic sealant

Absorbent filling

(b) $D/3$

Induced crack

Plastic crack inducer

Mastic sealed joint

2nd lane

1st lane

(c) $D/4$

12 mm tie bars 900 mm long at 750 mm spacing

Fig. 13.18 Concrete road construction joints (a) Expansion joint, (b) Contraction joint (c) Longitudinal joint

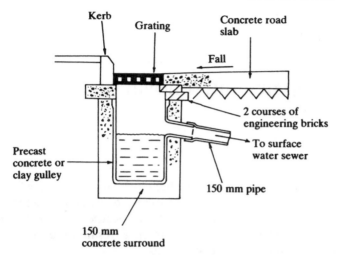

Fig. 13.19 Estate road drainage

14

Repairs and renovations

Repairs and remedial work to buildings are necessary when the design life of a component has expired, or when a component is subject to premature decay or failure because of poor design or installation. This may also occur as a result of using certain materials in incompatible situations.

This section is not concerned with the life span of correctly employed materials; it relates to structural or material defects occurring during the lifetime of a product or element of a building. Before considering individual elements it is worth exploring the potential structural damage to timber by woodboring insects and fungal attack, more commonly known as dry and wet rot.

Woodworm

Woodworm is a general term applied to all forms of wood-boring insect. The three most common wood borers in the British Isles are; the common furniture beetle, the death watch beetle and the lyctus or powder post beetle. There has also been a predominance of wood infestation from the house longhorn beetle in parts of south-east England.

In adult form these insects are beetles, although most of their life is as larvae, gnawing and burrowing through timber. Figure 14.1 shows a representative comparison of these creatures which also demonstrates the destructive potential of the much larger longhorn beetle and its 200 larvae of about 25 mm length and 6 mm diameter.

Life cycle

The life cycle is complex and divided into four successive stages; egg, larva (grub), pupa (chrysalis) and adult, (see Fig. 14.2). Eggs are laid in the rough crevices or former boreholes in the wood by

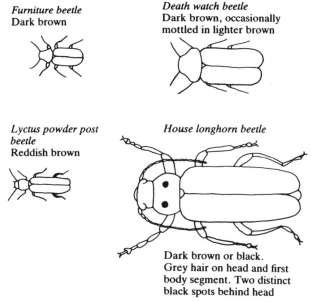

Furniture beetle
Dark brown

Death watch beetle
Dark brown, occasionally
mottled in lighter brown

*Lyctus powder post
beetle*
Reddish brown

House longhorn beetle

Dark brown or black.
Grey hair on head and first
body segment. Two distinct
black spots behind head

Fig. 14.1 Common wood borers. Note: Beetles are shown in adult
form in true proportion. House longhorn beetle is
approximately 25 mm long

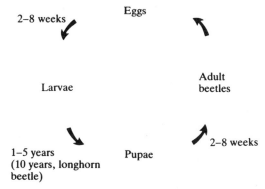

Eggs

2–8 weeks

Larvae

Adult
beetles

1–5 years
(10 years, longhorn
beetle)

Pupae

2–8 weeks

Note: Figures given are a generalisation, see Table 14.1 for more detail.

Fig. 14.2 Woodworm life cycle

adult female beetles. These hatch into larvae which bore into the
wood using it as food and shelter; in so doing, effecting
considerable damage particularly as each beetle produces a
multitude of larvae. After several years the larva changes to a

Table 14.1

Name	Egg quantity	Egg maturity	Laval period	Pupal stage	Emergence as beetle	Location	Exit holes
Furniture beetle	20–50	4–5 weeks	2 years	4–8 weeks	May–Sept.	Soft or hardwood, prefers softwood	1–2 mm diameter
Death watch beetle	40–80	2–8 weeks	4–5 years	2–4 weeks	Mar.–June	Hardwoods, preferably oak in damp situations	3 mm diameter
Lyctus (species vary)	70–200	2–3 weeks	1–2 years	3–4 weeks	May–Sept.	Sapwood of new hardwood	1–2 mm diameter
House Longhorn Beetle	Varies up to 200	2–3 weeks	Up to 10 years	About 3 weeks	July–Sept.	Sapwood of softwood	5 × 10 mm oval

Note: Periods given may reduce considerably in warm climates.

chrysalis just beneath the timber surface and emerge as an adult beetle to reproduce and generate more damage. This cycle varies slightly between the different species, therefore characteristics are compared in Table 14.1.

Recognition and treatment

Woodworm is usually first recognised by the distinctive flight holes and powdery deposits (frass) which appear on the timber surface. By this stage, the internal structure of the timber may have suffered considerable damage. The extent of damage may be established by chiselling away the timber surface to examine the borings or galleries produced by the larvae.

If the damage is slight and there are only a few exit holes, remedial spraying or brushing with liberal applications of a proprietary insecticide will control the problem. Structural timbers seriously damaged must be removed and burnt. All new timber must be pretreated, preferably with a pressure-impregnated insecticide, as it is extremely difficult to eradicate infestation completely. Also, new timber is likely to be low density sapwood, a very popular diet for the most abundant furniture beetle.

Prevention of woodworm infestation by use of pretreated timber is the only satisfactory method of control. The Building Regulations enforces the use of pretreated timber in roof spaces where construction is in parts of Surrey and adjacent areas. This is intended to counter the possibility of damage from the house longhorn bettle – once very predominant in Camberley, Surrey. The only satisfactory protection is extension of the regulations on a national scale to include all structural timbers, not just the roof structure. However, there is valid evidence to suggest that the toxic fumes from treatment chemicals could be harmful to health, hence the limit on roof timbers only. Acceptable methods of preservation are detailed in the Building Regulations and these include:

1 Diffusion with boron salts.
2 Dipping in organic solvents of chlorinated hydrocarbon insecticide.
3 Vacuum/pressure impregnation with copper/chrome/arsenic solution.
4 Steeping in creosote

Ventilation and prevention of dampness will also help to preserve timber from woodworm, as damp, draught free areas are popular breeding zones.

Dry and wet rot

Various fungal growths develop and feed on wood. The two most destructive and most commonly occurring are dry rot (*merulius lacrymans*) and wet rot or cellar fungus (*coniophora cerebella*).

Dry rot is the worse, as it can penetrate brickwork, plaster and concrete to establish itself on more timber. Furthermore, once established on damp timber it can spread and thrive on dry timber even if the original source of dampness is removed. It develops white fungal threads which conduct moisture from the timber (or damp atmosphere); these produce strands and spores (seeds) which drift in the air until they land and germinate on timber having a moisture content over 25%.

Wet rot is more restricted, it only develops in continuously damp conditions and will not spread to dry timber. The growth pattern is the same as dry rot except the fungal threads are usually dark brown or black. Thin olive-green or brown fruiting bodies are often the first indication of wet rot. These are clearly distinguishable from the flat grey and red fruit of the dry rot fungus.

Prevention

Most timber decay from fungal attack is caused by faulty constructions, e.g. lack of dpc; or general deterioration, e.g. broken roof tiles. New buildings can be affected by dry rot if structural components, particularly timber, are installed wet and the dampness is built in behind dry linings of plasterboard. It is essential to keep all stored structural timbers dry, at least with a plastic sheet covering during inclement weather. Air bricks in the external wall must be adequate (2 m max. spacing) and in effective positions (450 mm max. from quoins) to ensure sufficient ventilation under suspended timber floors to prevent condensation. Airbricks in roof gables or ventilated eaves construction are also essential means of air circulation through the roof structure, (see Chapter 6).

Treatment of dry rot

1   Eliminate source of dampness.
2   Dry out the affected area.
3   Inspect all timber in the vicinity of attack.
4   Remove and burn all fungoid timber and sound timber within at least 500 mm of attack.

5   Chisel off contaminated plaster and rake out brick joints behind.

6   Sterilize surfaces of concrete or brickwork in contact with the contamination. This is achieved by heating with a blow lamp until the surface is too hot to touch, followed by generous use of a liquid fungicide applied as the surface cools.

7   Deeply impermeated walls must be treated by irrigation – holes bored at 500 to 900 mm centres and fungicide pressure impregnated through the holes and into the thickness of the wall.

Fungicides are supplied in conveniently sized cans as water diluted formulations of sodium pentachlorophenate, sodium orthophenylphate or possibly mercuric chloride. All preservatives can be harmful to the skin, therefore protective gloves and goggles are advisable during application. Replastering should be preceded by a cement and sand rendering containing a zinc oxychloride fungicide.

Treatment of wet rot

1–4 as dry rot.

5   If the fungus can be identified as wet rot, sterilization of adjacent surfaces is unnecessary. Drying is sufficient to prevent further damage.

Whether dry or wet rot, all replacement timbers should be pressure impregnated with preservative. Ventilation should be checked as adequate or improved, and installation of central heating will also assist in preventing further attack.

Rising damp

This potentially damaging phenomena is often responsible for the dry- and wet-rot problems just explained. If water is to rise in a wall a constant supply must be available at the base. Water rises by an upward capillary pull between the masonry pores. The smaller the pores the higher water is likely to rise, although heights in excess of 1 m are unlikely as the mass of water exceeds the wall suction potential and much is lost by evaporation. The most obvious indication of rising damp is dark staining above the skirting board on the interior of a wall. This must not be confused with dampness from other causes, which could include:

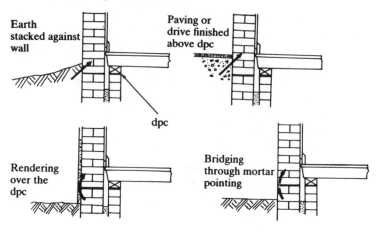

Fig. 14.3 Bridging damp proof course in solid walls

1   Condensation
2   Defective plumbing
3   Leaking gutters and downpipes
4   Defective chimney/flashing
5   Saturation through solid wall construction.

Efflorescence is another indicator, appearing as white salty deposits on both faces of a wall. These salts are drawn from the ground as the dampness rises and they combine with any salts in the masonry. Apart from being unslightly, these salts have hygroscopic properties and will absorb moisture from the air, restricting water evaporation from the wall.

Rising damp may be attributed to several possible constructional faults as shown in Fig. 14.3 for solid walls and Fig. 14.4 for cavity walls. Measurement of the surface moisture is initially with a battery powered electric meter with two pointed electrodes inserted into the wall. The meter reads the resistance between the two electrodes and presents this as a moisture content on the meter gauge. As a preliminary indicator, these meters are useful for identifying areas which require further investigation. However, they do not necessarily identify a rising damp problem, as often positive meter readings are obtained from virtually dry walls where there is a high concentration of surface salts. True evidence of rising damp can only be obtained from samples taken from borings into the centre of a wall.

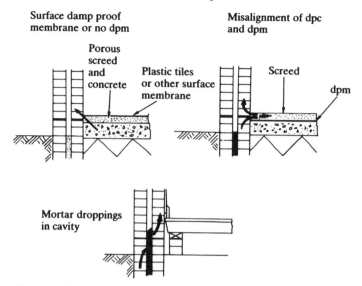

Fig. 14.4 Bridging damp proof course in cavity walls

Treatment of rising damp

After removing any obvious bridges to the damp-proof course, such as earth piled against the external wall and rendering over the dpc, internal plastering is removed to a height of 1 m to eliminate the possibility of further efflorescence. The installation of land drainage with perforated or porous pipes may also be considered adjacent to the base of the wall, if the area is permanently saturated from a high water table.

There are several proprietory techniques which purport to arrest the problem of rising damp, but only chemical injection has attained proven standards acceptable for issue of an agrément certificate. Another successful method is physical insertion of waterproof material in the full width of a wall after forming a horizontal slot with a tungsten-carbide tipped saw.

Chemical injection system

Holes of 12 mm diameter are bored into both sides of the defective wall at 100 mm intervals approximately 150 mm above ground level. Solutions of silicone or aluminium stearates are gravity transfused (see Fig. 14.5) or high-pressure injected until full saturation of the wall is achieved. Holes are made good with cement and sand mortar, rammed into the full depth with a piece

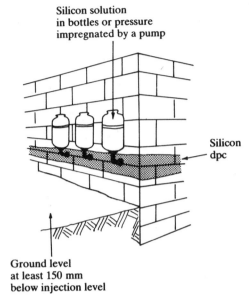

Silicon solution
in bottles or pressure
impregnated by a pump

Silicon dpc

Ground level
at least 150 mm
below injection level

Fig. 14.5 Gravity transfused silicon injection

of timber doweling. A complete house can be treated in two to three days, and most reputable installers offer a 30-year guarantee against further rising damp. The physical effect is chemical bonding to, or lining of the pores by curing and solvent evaporation. This is sufficient to prevent the rise of liquid moisture, but may still permit passage of some water vapour. Therefore, the success of this system is mainly in controlling the problem rather than by provision of a complete damp proof barrier. The small amount of water that penetrates as vapour is lost by evaporation. The optimum time for installation is during the late summer months when the water table is lowest. This particularly applies if gravity transfusion is used as it could be unsuccessful if the wall is heavily saturated.

Physical damp-proof course

This is the only fully effective remedial treatment for a defective or non-existent damp-proof course. It is a slow tedious process involving cutting out a complete horizontal course of mortar in about 1 m lengths, with a tungsten-carbide tipped chainsaw. The new dpc of polythene, bituminous felt or non-ferrous metals, e.g.

copper or lead is loaded with mortar and inserted into the void. 100 mm should be allowed for overlapping at each end. It is essential to ensure full penetration of the dpc to the interior of the wall and temporary wedges should be positioned to prevent superstructural settlement. Great care must be taken to avoid service pipes and cables. If necessary the new dpc can extend into the building and overlap a polythene damp proof membrane in a concrete floor. The major disadvantage is the high professional cost, incurred as a result of the slow but nevertheless thorough process. However, it may appeal to someone of semi-skilled capabilities, undertaking a personal restoration project, as the special bladed chain saws are hireable and the old mortar joints will probably be of soft lime mortar composition. Another disadvantage is its unsuitability to uncoursed or random stone walls, often associated with aged property in need of remedial treatment.

Superficial faults

Plaster

Most faults and defects associated with plasterwork relate to:

1　Poor background preparation.
2　Poor workmanship (application).
3　Defective plaster (stored too long).
4　Uncontrolled drying.

Common faults

1. Blistering. Too much radiant heat on the finishing coat causes it to lift and lose its adhesion. Likely to occur during the winter months when artificial heat sources are used to accelerate the setting process.

2. Bond failure. This is noticeable where large areas of plaster sound hollow when tapped. It is more obvious where patches fall away from the wall. Causes are usually associated with impurities on the background wall, e.g. soot staining from old unlined flues, or areas of unset plaster produced by premature drying. Other causes are salts or efflorescence on the background and undercoat interface or inadequate preparation of the background which should be hacked if smooth and dampened to encourage adhesion.

3. Cracking. This is due to structural shrinkage, and is particularly noticeable in new buildings which dry out slowly with occupation. Areas particularly affected are internal wall corners and the junction between plasterboard ceiling and wall. Some minor cracking is often noticeable around door and window lintels and joinery frames shrink and adjacent walls settle. Elongated cracking in the horizontal or vertical planes is more serious and generally associated with excessive foundation settlement.

4. Crazing. These are fine cracks, often found in isolated areas. They arise from outer surface shrinkage, which could also cause loss of bond. The cause is either a finishing coat applied too thinly or the use of a 'stale' finishing grade plaster from a bag which has been open too long.

5. Efflorescence. This is salt contained in the background brickwork or mortar, surfacing and drying out. Possibly also attributed to impurities in the plaster.

6. Grinning. More appropriate to rendered walls, where a thin layer shows the background mortar joints 'grinning' through in damp weather. Can apply to thin plaster applications where the different suction effects of wall material and mortar show through at certain angles of vision.

7. Popping or blowing. Air bubbles surface as the plaster is applied to spoil the finish with tiny craters. The cause is impurities in the plaster, probably acquired during mixing on a dirty surface or spot board.

8. Rust stains. This happens where plaster is applied to rusty metal lathing. All lathing should be galvanised to a high standard, but bad site handling may cause the surface to break down and rust in the presence of wet plaster.

9. Soft patches or chalkiness. This is a powdery surface defect caused by the plasterer working the finish coat past its setting point. It also occurs when the plaster sets too quickly as a result of excessive heat – possibly from artificial driers.

10. Surface dampness. Often associated with rising damp or where rising damp has been treated. It is due to salts left in the wall or possibly from the sand in the rendered background, attracting dampness from the atmosphere.

11. Surface irregularities. Poorly aligned brick or block background wall. Certain irregularities can be masked with a generous application of rendering, or possibly two rendered undercoats.

Paint

Faults in painted surfaces often arise in consequence to prolonged storage before use, storage in varying temperatures, use of the wrong thinner and moisture contamination which is often related to acidity in metal pigmented paints.

Common faults occurring after application include:

1. Bloom or surface haze. This is a greyish white surface discolouration caused by moisture contamination during drying. Condensation from a humid atmosphere and lack of ventilation can be responsible for this defect as well as dampness from rainfall. Cellulose lacquers are particularly prone to this problem.

2. Blistering. A very common fault on newly painted joinery which has been badly stored on site during construction. It is caused by vaporisation of moisture in the timber failing to escape through the paint barrier. Blisters eventually crack and allow the moisture out, but the paint peels away. It can also occur by absorption of water through an inadequately prepared end grain. Sills are particularly prone.

On metals a similar problem results from insufficient surface preparation. This is usually related to rust or absence of the correct primer. Electrolytic corrosion due to connection of dissimilar metals will also cause blistering of paintwork.

3. Cracking. This usually appears in fairly regular patterns, hence the terms, 'checking' or 'aligatoring' often used to describe this problem. It is caused by applying the finish coat before the undercoat is completely dry, and the differential drying between the two layers creates shrinkage cracks on the surface.

4. Peeling and flaking. Where complete areas or strips of paint can easily be peeled off, the fault is frequently a dirty or contaminated surface. Formwork mould oil or diesel oil inadvertently spilt on pre-fixed window or door frames could be a cause, or possibly use of an incorrect surface primer.

5. *Poor gloss.* In the absence of dampness, this will occur on exceptionally porous surfaces or more likely, where the priming and undercoats are overthinned or too few.

6. *Slow drying.* This leads to an excessive period of tackiness which attracts insects, cement powder, dust, etc to the surface. The causes may be moisture contamination, low temperatures and use of old paint which has suffered some evaporation of the drying agent.

7. *Wrinkling.* Wrinkles in the paint surface are caused by too much paint being applied in one layer. As the surface dries, it folds into the soft under layer. This is mainly true for horizontal surfaces; in vertical or inclined surfaces the paint simply runs.

Construction defects

Defects and faults in construction may arise in any of the areas considered in this book. If practice is in accordance with the recommendations herein, with specific reference to Building Regulations, British Standards, Codes of Practice, Agrement certificates and NHBC rules, problems should not arise. However, the human factor – workmanship – is dependent on proven standards of competence and adequate supervision. A high degree of competency and supervisory skills have unfortunately not always benefited the building industry, but now government training schemes, apprenticeships, professional short courses, etc. are responsible for a well-educated and efficient team. Advances in material testing and development have also contributed to higher standards throughout the industry.

As a check on avoiding constructional defects with reference to current legislation and codes of practice the reader is recommended to refer to the following supplementary information: Building Research Establishment Digest No. 268 – Common defects in low-rise traditional housing.

Procedure during a structural survey of an existing dwelling

A structural survey should be presented as a bound document

containing clear concise information about every aspect of the construction, (subject to client's requirements). Where appropriate, illustrations and photographs should accompany the report.

A preliminary section states:

- Client's name and address
- Address of property
- Date of survey
- Whether property is occupied or empty
- Whether freehold or leasehold
- Purpose or extent of survey (varies depending on client's requirements)
- Whether specialists are attending, eg. electrician or heating engineer
- Age of property
- Accommodation
- Elevation and position
- Area of estate
- Access, pedestrian and vehicle

Before commencing the survey, the following equipment should be checked:

- Notebook and pencil
- Camera (with flash)
- Binoculars (chimney inspection)
- Tape measure
- Spirit level and/or plumb line
- Torch
- Screwdriver (floor-board access and for probing timber and pointing)
- Ladder (access to eaves)
- Manhole cover lifting irons
- Moisture meter
- Hammer

Order of survey

Most surveyors prefer to commence on the outside, starting at high level and working down, followed by an interior examination. The following sections are for guidance only, they are not intended as the necessary order of progress:

Roof

(a) *Exterior*

- Type of tile ⎫
  Hips ⎪
  Ridge ⎬ treatment
  Verge ⎪
  Valley ⎭
- General condition and defects
- Eaves; type of construction and defects
- Gutters; material, size, condition
- Rainwater goods; ditto
- Chimney; verticality, pointing, flashing
- Abutments; flashing and weathering

(b) *Interior*

- Accessibility
- Type of construction, i.e. trusses or traditional
- Evidence of woodworm/dry or wet rot
- Water penetration around chimney or valleys
- Insulation; adequacy if existing
- Cold-water storage cisterns, condition and suitability, whether insulated
- Pipework; material, condition and insulation. Test overflows
- Ventilation

Walls
- Material, condition
- Pointing
- Air bricks; location, suitability and efficiency
- Damp-proof course; material, condition, effectiveness, location relative to ground level
- Windows; type, material, condition, quality of paintwork and glazing
- Doors; type, material, condition particularly threshold
- Location of frame to wall – effectiveness
- Evidence of settlement – if positive check foundations and subsoil for movement

Drainage
- Inspection chambers; access covers, condition of channels,

benching and general interior – penetration of tree roots
- Pipes; materials and capacity
- Tests; air pressure test between inspection chambers and stack. Measure flow rate, (self cleansing velocity)
- Gulleys, type, condition and size
- Cesspools and septic tanks; volume, structural condition and function – check surface crust in septic tank and observe outfall if possible
- Sanitation; soil, waste and ventilating pipe – size of stack and branches, gradient on waste branches, performance test to check seal retention in traps

External works
- Establishment of boundaries
- Fences and gates; material, condition adequacy
- Hedges
- Trees; proximity to building, height if relative
- Paths and drives; condition and suitability
- Outbuildings; garage, barn, sheds, greenhouses, etc. general condition
- Ordnance datum level; possibility of flooding, nearby water courses.

Floors
- Ground floor; construction, finish, stability. If timber examine for rot below surface, clear ventilated space, dpc on sleeper walls
- Upper floor; likely to be timber, check for woodworm.
- Stability relative to wall fixing and span to size/loading

Stairs
- Construction; pre formed or built *in-situ*
- Soffit; plasterboard?
- Tread condition
- Balustrade; suitability and stability
- Safety; defects, pitch angle, open tread, handrail

Interior finishes
- Decorations; wall paper, faded, torn, defective
- Woodwork; condition, defects, paintwork
- Plaster; plasterboard ceiling – fixing, general finish and condition

- Fittings; iron mongery, whether damaged, sufficiently secured to prevent draught, security

Services
- *Water*; service pipe isolation valve, position and effectiveness
- Stop and drain cocks
- Rising main, material and position. Insulation – if necessary
- Control, position of valves, accessibility
- Cold-water supply from cistern; control, material, sufficient supply
- Hot-water supply; control, capacity and adequacy of storage, material, insulation, condition including storage cylinder
- Specialists report
- *Electricity*; controls, fuses, consumer service unit. Ring main, lighting circuits – efficiency of earth leakage
- Materials, cable sheatting
- Date of installation
- Immersion heater – function
- Specialist's report
- *Gas*; installation pipework size and material
- Test for leakage
- Effectiveness of individual appliances
- Specialist's report
- *Central heating*; effectiveness, efficiency of controls, thermostats etc.
- Specialists report

INDEX

fork lift 35
formwork clamp 146–7
formwork, column 145–6
formwork, floor 115, 118
formwork, stair 215–7
foundation design 50–4
foundations 55–65
frame 139, 184
frame tie 97, 186
French drain 45–6
frieze rail 190
fresh-air inlet 271
frontage line 37–9
frost heave 45, 55
furniture beetle 294–7

gable 122
gable ladder 133
ganger 11
gate valve 253, 255
gauge 134, 137–8
gauged arch 95
glass 203
glazed wall tile 235–6
glazing 195, 203
glazing bar 198
glazing bead 204
glue block 213, 215
grab crane 34
gravel driveway 290
ground floor 104, 113
ground water control 44
guard pipe 248
guard rail 101–3
gulley 265–8, 293
gusset 164
gymnasium floor 225

half-brick wall 79
half-round channel 270
handrail 209, 213
hanger 128–9
hardcore 113
hardwood floor 223–5
hearth 170–2
heating systems 256–8
herring bone strutting 106–7
hip 122
hip iron 135
hoarding 26, 28–9
holing and notching 108–9
hollow beam 119
honeycombe wall 105
hook bolt 165–6
hot water storage cylinder 251, 254

hot water supply 254–6
housed joint 107–8

I beam 119–20
impact sound 112
imposed load 50
independent scaffold 102
indirect cold water 249–51
indirect hot water 254–5
infill panel 139, 147
inspection chamber 265–8
insulated cavity wall 58, 61, 62, 83, 84, 95
insulated ground floor 96, 105, 114
insulated upper floor 112
insulation quilt 112
insulation, roof 138–9, 195
interceptor trap 271
inverted channel 119
ironmongery 205–7

jamb 97
JCT contract 7
jointing
  bricks 72
  clay drainpipes 259–61
  plastic drainpipes 259–61
joist hanger 82, 107–8, 125
joist design 108–12
joists 104–12

kerb 291–3
kicker 140
king post 131
kite winder 211

labour 17
land drains 45–9
lateral restraint 81–2
lath and plaster 231
lathing 168, 229
lattice beam 154–5
lattice frame 155, 162
lead flashing 125–6, 179–83
lead sheet tools 183
lead valley 135
lean-to roof 122–6
ledged and braced door 187
ledger 101–3, 117
levelling 31, 36, 38
lining 91, 94, 184–6, 195, 199
links 145–6
lintels 92–4, 98–100
local authority 5, 17, 27
longhorn beetle 294–6